DNA SEQUENCING

from experimental methods to bioinformatics

The INTRODUCTION TO BIOTECHNIQUES series

CENTRIFUGATION
RADIOISOTOPES
LIGHT MICROSCOPY
ANIMAL CELL CULTURE
GEL ELECTROPHORESIS: PROTEINS
PCR, SECOND EDITION
MICROBIAL CULTURE
ANTIBODY TECHNOLOGY
GENE TECHNOLOGY
LIPID ANALYSIS
GEL ELECTROPHORESIS: NUCLEIC ACIDS
LIGHT SPECTROSCOPY
DNA SEQUENCING

Forthcoming titles

MEMBRANE ANALYSIS
PLANT CELL CULTURE

DNA SEQUENCING

from experimental methods to bioinformatics

Luke Alphey

School of Biological Sciences, The University of Manchester, Manchester, UK

Luke Alphey
*School of Biological Sciences, The Universty of Manchester,
Manchester, UK*

**Published in the United States of America, its dependent territories
and Canada by arrangement with BIOS Scientific Publishers Ltd,
9 Newtec Place, Magdalen Road, Oxford OX4 1RE, UK**

A CIP catalogue record for this book is available from the British Library.

Library of Congress Cataloging-in-Publication Data
Alphey, Luke.
  DNA sequencing / Luke Alphey.
    p. cm. -- (Introduction to biotechniques)
  Includes bibliographical references and index.
  ISBN 0-387-91509-5 (hardcover : alk. paper)
  1. Nucleotide sequence. I. Title. II. Series: Introduction to
  biotechniques series.
  QP625.N89A45 1997
  572.8′633--dc21   97-20709
                                   CIP

ISBN 0-387-91509-5 Springer-Verlag New York Berlin Heidelberg
SPIN 19900581

Springer-Verlag New York Inc.
175 Fifth Avenue, New York
NY 10010-7858, USA

Production Editor: Priscilla Goldby
Typeset by Chandos Electronic Publishing, Stanton Harcourt, UK.
Printed by Biddles Ltd, Guildford, UK.

Contents

---

8.   Troubleshooting   **81**

PART 2: APPLICATIONS

---

9.   Confirmatory Sequencing   **89**

---

10.   Sequencing PCR Products   **97**

Abbreviations

BLAST	Basic Local Alignment Search Tool
BSA	bovine serum albumin
CASP	critical assessment of structure prediction
CCD	charge-coupled device
cDNA	complementary DNA
DDBJ	DNA Databank of Japan
DDGE	double-strand denaturing gel electrophoresis
ddNTP	2′, 3′-dideoxynucleotide
DMSO	dimethylsulfoxide
DNA	deoxyribonucleic acid
DTT	dithiothreitol
EBI	European Bioinformatics Institute
EDTA	ethylenediamine tetraacetic acid
EMBL	European Molecular Biology Laboratory
EPD	Eukaryotic Promoter Database
EST	expressed sequence tag
ExoIII	exonuclease III
ftp	file transfer protocol
GCG	Genetics Computing Group
HPLC	high-performance liquid chromatography
HSSP	homology-derived structures of proteins
IUB	International Union of Biochemistry
IUPAC	International Union of Pure and Applied Chemistry
MCS	multiple cloning site
5-MeC	5-methylcytosine
mRNA	messenger RNA
NCBI	National Center for Biotechnology Information
NMR	nuclear magnetic resonance
NP-40	Nonidet P-40
OMIM	Online Mendelian Inheritance in Man
ORF	open reading frame
PC	personal computer
PCR	polymerase chain reaction
PDB	protein databank
PEG	polyethyleneglycol
PNK	poynucleotide kinase
REBASE	Restriction Enzyme Database
RFLP	restriction fragment length polymorphism
RNA	ribonucleic acid
RT–PCR	reverse transcriptase–polymerase chain reaction
SCOP	structural classification of proteins
SRS	Sequence Retrieval System

SSCP	single-strand conformation polymorphism
STS	sequence tagged site
TBE	Tris–borate–EDTA
TE	Tris-EDTA
TEMED	N,N,N′,N′-tetramethylethylenediamine
TES	2-[Tris(hydroxymethyl)methylamino]-1-ethanesulfonic acid
TREMBL	Translated EMBL
TTE	Tris–taurine–EDTA

Preface

In the 20 years since the current methods were first introduced, DNA sequencing has been at the heart of modern molecular biology. The sequence databases have been growing at an exponential rate, and even that rate of increase is improving, with doubling time down from about 22 months to 9 months. Whole new areas of research have been opened up by this technology, from molecular genetics to molecular taxonomy. With the advent of whole genome sequencing, exciting new vistas are emerging.

This book is intended as a practical guide, particularly at the strategic level. It aims to explain the options available and their relative merits, to allow the reader to decide which is most suitable for their application. The book covers the whole process of DNA sequencing, from planning the approach, through data acquisition, to extracting useful biological information from the data.

The book is aimed primarily at those new to DNA sequencing, but I hope that it will also prove a useful text for more experienced sequencers and that the information provided will be useful as a source of further information on familiar techniques and as a reference for less common ones. Part 1 describes the basic methods in detail, including manual and automated sequencing and the various pitfalls that may be encountered on the way. The equipment required is discussed, together with the advantages and disadvantages of each option.

Part 2 details the major applications of DNA sequencing: confirmatory sequencing to check a particular construct or mutant; sequencing PCR products; and strategies for sequencing large fragments of uncharacterized DNA. Part 3 covers Bioinformatics – the analysis of the sequence data to extract useful information. This section was contributed by Dr Andy Brass, Senior Lecturer in Bioinformatics at the University of Manchester, UK. It covers sequence analysis from checking and compiling the raw data through to homology searches and structural predictions.

Luke Alphey

Acknowledgements

First of all, I would like to thank Andy Brass for contributing the Bioinformatics section of the book. Jane Hewitt provided most of the gel examples for *Table 8.1* and Lawrence Hall provided the data for *Figure 7.2*. Eaton Publishing (*Figure 7.4*), VCH Verlagsgesellschaft mbH (*Figure 7.6*) and PE-Applied Biosystems (*Figures 10.3–10.6* and *10.8*) all generously permitted the reproduction of their copyright material. I am also grateful to Jane Hewitt, Lawrence Hall and Nina Nicholls for their critical reading of the manuscript. Finally, I would like to thank N.N. and B.B. for their constant encouragement and support.

Safety

Certain reagents indicated for use in this book are chemically hazardous or radioactive. The researcher is cautioned to exercise care with these reagents and with the equipment (e.g. electrophoresis equipment) used in these procedures, strictly following the manufacturer's safety recommendations. Disposal of waste (including waste chemicals and radioactive materials), must comply with all local, national and other applicable regulations. These procedures may also be governed by other relevant regulations, for example those covering the containment and use of genetically modified micro-organisms. While every care has been taken to ensure that the experimental details discussed in this book are accurate and safe, the author accepts no liability for any loss or injury howsoever caused.

Many of the procedures discussed in this book are protected by patents or other legal protection. The reader is hereby notified that the purchase of this book does not convey any license or authorization to practise any of these procedures.

1   What is DNA Sequencing?

1.1 An introduction

DNA sequencing is the determination of all or part of the nucleotide sequence of a specific deoxyribonucleic acid (DNA) molecule. The ability to sequence DNA lies at the heart of the molecular biology revolution. Techniques to sequence DNA were developed only quite recently; the original papers describing the modern methods were published in 1977 [1, 2]. The rate at which new sequence information is determined has increased rapidly over the last 20 years. It is still accelerating, to the extent that the entire human genome sequence of approximately 3×10^9 base pairs will be determined within the next few years, as will the genome sequences of a considerable number of other organisms of medical, agricultural or scientific importance.

The fundamental reasons for wishing to know the sequence of a DNA molecule are:

- to make predictions about its function;
- to facilitate manipulation of the molecule.

The aim of this book is to show how DNA sequence information is obtained and analyzed, and some of the major reasons for doing so. Chapters 2–8 describe the sequencing methods in common usage, with particular emphasis on the relative merits and pitfalls of each approach. Chapters 9–11 describe the major applications of these methods. Subsequent chapters cover the computer-based analysis of sequence data.

Before discussing the principles behind DNA sequencing, we must first consider the structure of a DNA molecule.

1.2 Nucleic acid structure

The normal conformation of DNA is as a double helix (see *Figure 1.1*). This helix comprises two DNA strands running antiparallel to each other, each strand being a chain of bases, each base covalently linked to the next. The bases are each attached to deoxyribose, a sugar molecule, and each sugar molecule is linked to the adjacent sugar molecule via a phosphate group. The basic repeat unit of DNA therefore comprises a base, a sugar and a phosphate group, and is known as a nucleotide (see *Figures 1.2–1.5*).

The structure of a four-nucleotide segment of DNA is shown in *Figure 1.6*. Note that only one strand is shown. Note also the numbering of the carbon atoms in the deoxyribose (sugar) part of the molecule. These each have a 'prime', for example 5′ and 3′, to distinguish them from the atoms of the bases. It is the 5′ and 3′ carbons of adjacent sugars that are linked via the phosphate groups, so each covalently linked DNA strand will have a 5′ end and a 3′ end, as shown in

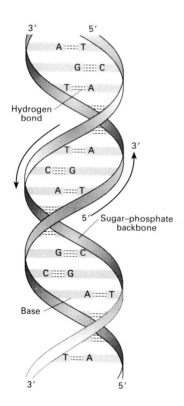

FIGURE 1.1: *The DNA double helix. Reproduced from Williams* et al. *(1993)* Genetic Engineering, *BIOS Scientific Publishers Ltd.*

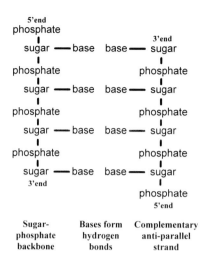

FIGURE 1.2: *Components of a DNA helix. A single strand of nucleic acid has a sugar–phosphate backbone to which the bases are attached. These linkages are all covalent. The other strand runs antiparallel. The two strands are held together by hydrogen bonds formed between complementary bases.*

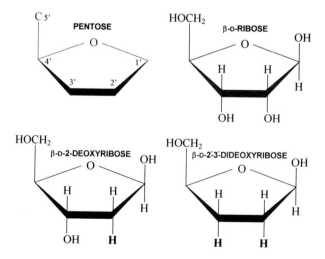

FIGURE 1.3: *Sugar structures of rNTPs and ddNTPs. The sugar–phosphate backbone of RNA contains the 5-carbon sugar ribose, whereas that of DNA contains 2′-deoxyribose. ddNTPs which are used in DNA sequencing by the chain termination method (Chapter 3) contain the synthetic analog 2′, 3′-dideoxyribose. The standard numbering system for the carbon atoms is shown for a generic 5-carbon sugar (pentose). Carbon atoms in the sugar part of nucleotides are designated 1′, 2′, etc. to distinguish them from the atoms in the base.*

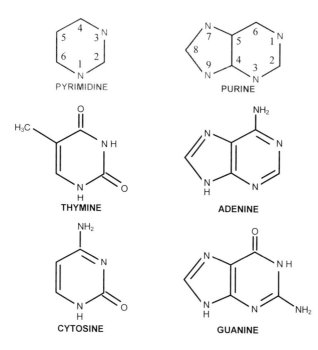

FIGURE 1.4: *The structure of the bases found in DNA. Thymine and cytosine are pyrimidines, adenine and guanine are purines. The numbering system for the ring atoms is shown. Pyrimidines are linked to the sugar at N_1, purines at N_9.*

Figure 1.6. In a linear double-stranded molecule, the 5′ end of one strand is complementary to the 3′ end of the other strand.

The two strands of the double helix are held together noncovalently by hydrogen bonds. The hydrogen bonds form between the complementary bases: adenine (A) pairs with thymine (T) and guanine (G) with cytosine (C) (see *Figure 1.7*).

Nucleoside = base + sugar

	base	nucleoside
A	adenine	adenosine
C	cytosine	cytidine
G	guanine	guanosine
T	thymine	thymidine
U	uracil	uridine

Nucleotide = base + sugar + phosphate

ATP	adenosine triphosphate
dATP	deoxyadenosine triphosphate
ddATP	dideoxyadenosine triphosphate
dGTP	deoxyguanosine triphosphate
dNTP	deoxynucleoside triphosphate
ddNTP	dideoxynucleoside triphosphate

FIGURE 1.5: *Nomenclature of nucleic acid precursors. The abbreviations A, C, G, T, etc. are usually clearer than the full names. dNTP and ddNTP can contain any base and are often used to refer to an equimolar mix of all four (di)deoxynucleoside triphosphates.*

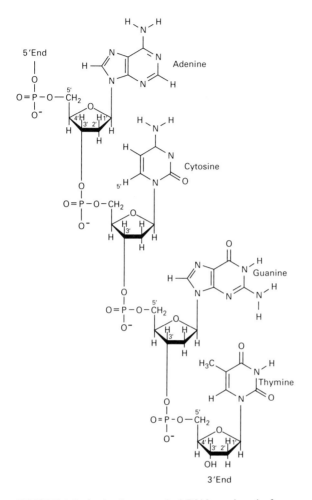

FIGURE 1.6: *A single-stranded DNA molecule four nucleotides in length. Reproduced from Newton and Graham (1994)* PCR, *BIOS Scientific Publishers Ltd.*

The related nucleic acid RNA (ribonucleic acid) differs from DNA in that the sugar in the sugar–phosphate backbone is ribose, rather than deoxyribose (see *Figure 1.3*), and uracil (U) is used in place of T.

1.3 DNA sequencing

Methods for sequencing RNA were developed earlier than for DNA, but now RNA is rarely sequenced directly. Instead, a complementary

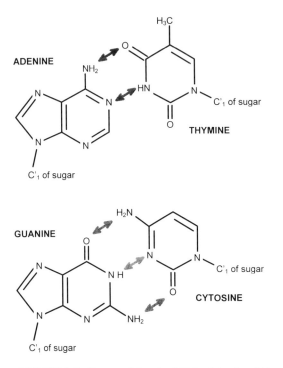

FIGURE 1.7: *Base pairing in DNA. Adenine (A) pairs with its complementary base thymine (T) and guanine (G) with cytosine (C). In RNA, uracil (U), replaces thymine. Note that the separation between the glycosidic bonds and the sugars are exactly the same (10.85 Å) for each base pair.*

DNA (cDNA) copy is synthesized. This cDNA is then sequenced, and the sequence of the original RNA deduced from this.

DNA sequencing is the determination of the base sequence of all or part of a DNA molecule. In the case of the molecule shown in *Figure 1.6*, DNA sequencing would advance our knowledge from 'a DNA fragment about four bases long' to 'a DNA fragment whose sequence is ACGT' (see *Figure 1.8*). Of course most, DNA molecules of biological interest are considerably longer than this!

The informational content of DNA is encoded in the order of the bases (A, C, G and T) in much the same way as binary information is stored in a computer as a string of 1s and 0s (*Figure 1.9*). The purpose of DNA sequencing is to determine the order (sequence) of these bases in a given DNA molecule. However, knowing the DNA sequence of a gene does not necessarily tell us what that gene does, any more than knowing the binary code of a computer program will necessarily tell

Since DNA comprises only four different bases, which are joined covalently in a linear, unbranched sequence, we can conveniently represent the molecule shown in *Figure 1.6* as:

5′-ACGT-3′

DNA sequences are written 5′ to 3′ by convention so this is often written simply as:

ACGT

A base-pairs with T and G with C, so if this represents double-stranded DNA then we know that the double-stranded molecule is:

5′-ACGT-3′
3′-TGCA-5′

Note the reverse polarity of the complementary strand. In this example, the sequence of the two strands is in fact identical, but this need not be the case.

FIGURE 1.8: *Conventional notation for DNA sequences.*

us what the program does. In either case, we will normally need a great deal of additional information about the biological or electronic system which interprets the information.

There are two sequencing techniques in current use. These are the 'chain termination' and 'chemical degradation' methods, also known as the 'Sanger' and 'Maxam and Gilbert' methods respectively, after their original inventors. The chain termination method depends on

Let A=0, C=1, G=2, T=3

Then	5′	**A**	**C**	**G**	**T** 3′
is:		0	1	2	3
or in binary:		00	01	10	11

even base-pairing is easily modelled this way - the sum of base and complement is 3:

	5′	**A**	**C**	**G**	**T** 3′
		0	1	2	3
		00	01	10	11
complement:		11	10	01	00
		3	2	1	0
	3′	**T**	**G**	**C**	**A** 5′

FIGURE 1.9: *DNA stores digital information.*

the enzymatic synthesis of labeled DNA, using special modified nucleotides called dideoxynucleotides to terminate the elongating strand. The chemical degradation method is based on the base-specific chemical degradation of a DNA molecule labeled at one end. Although there are specific applications which use the chemical degradation method, the vast majority of DNA sequence today is determined by use of dideoxy chain termination.

Both methods generate a nested set of DNA molecules in which the length of each molecule represents the distance from a fixed point to a specific base. The sequence information is obtained by measuring the relative sizes of these molecules. Both methods therefore require the accurate separation of labeled DNA molecules which differ in size by only a single nucleotide. Sequencing 500 bases in a single reaction demands the accurate separation of molecules 499 bases long from those 500 bases long, a 0.2% size difference. This requires high-resolution denaturing gel electrophoresis, the resolution of which is often the limiting factor in determining how much sequence information can be obtained from a single reaction.

The base sequence is just a part of the complete description of the structure of a DNA molecule. For a start, some of the bases are modified *in vivo*, for example by methylation. Patterns of methylation vary from one organism to another. In humans, the majority of genomic DNA is methylated at CpG dinucleotides. About 2% of the genome is unmethylated, GC-rich regions ('CpG islands'). These are of interest to molecular biologists as they mark the 5′ ends of genes. In contrast, the fruit fly *Drosophila melanogaster* does not seem to methylate its DNA at all. Prokaryotes methylate DNA as part of 'restriction' systems which help to protect their genetic information ('genome') from contamination by foreign DNA. Some base modifications can be detected by current sequencing techniques, others can be probed by other sequence-specific methods, such as restriction endonuclease digestion (see Section 9.4).

RNA molecules can also contain modified bases and may contain various other sequences not directly complementary to the DNA molecule from which they were copied, for example the poly(A) tail found on mature eukaryotic mRNAs. As well as the primary structure (nucleotide sequence), the secondary structure of an RNA molecule may also be important for its function. The minimum-energy state of a nucleic acid is normally when hydrogen bonding between complementary bases ('base-pairing') is maximized. For a single-stranded molecule, this means that self-complementary regions will base-pair to form 'hairpin' structures, with a double-stranded stem and a single-stranded loop. These structures often affect protein binding, for example in translational regulation. Intermolecular base-pairing is also important, for example in codon–anticodon recognition during protein synthesis. DNA is predominantly double-stranded, but secondary structure formation of single-stranded DNA is a major cause of problems in DNA sequencing (see Sections 8.2 and 8.3).

In addition to restriction endonucleases, many other nucleic acid-binding proteins show at least some sequence specificity. These include the transcription and translation machinery responsible for interpreting the protein-coding information encoded in the DNA. The elements of this system are themselves proteins and RNA molecules, encoded by the DNA. The ultimate challenge of genome sequencing and analysis is to understand how this incredibly complex self-regulating and self-reproducing system describes and encodes a living organism.

References

1. Maxam, A.M. and Gilbert, W. (1977) A new method for sequencing DNA. *Proc. Natl Acad. Sci. USA,* **74**, 560–564.
2. Sanger, F., Nicklen, S. and Coulson, A.R. (1977) DNA sequencing with chain-terminating inhibitors. *Proc. Natl Acad. Sci. USA*, **74**, 5463–5467.

2 Chemical Degradation (Maxam and Gilbert) Method

2.1 A description of the method

The chemical degradation method uses the base-specific chemical cleavage of an end-labeled DNA molecule to generate a nested set of labeled molecules, each terminating at a specific base [1, 2]. Following high-resolution denaturing gel electrophoresis (Chapter 6) and detection of the labeled fragments, typically by autoradiography, the sequence of the original DNA can be read from the resulting sequencing 'ladder', just as in the chain termination method (see Chapter 3).

The base-specific cleavage is performed by using reagents that modify a specific base or bases (i.e. G, G+A, C+T or C) in such a way that subsequent treatment with hot piperidine cleaves the sugar–phosphate backbone at the sites of modification (*Table 2.1*).

TABLE 2.1: Base-specific reactions for chemical degradation method

Reagent	Base(s) affected
Dimethyl sulfate pH 8.0	G
Piperidine formate pH 2.0	A+G
Hydrazine	C+T
Hydrazine + 1.5 M NaCl	C
Hot (90°C) 1.2 M NaOH	A>C

These reagents react with DNA to modify specific bases. Hot piperidine (1 M in water, 90°C) then cleaves the sugar–phosphate backbone of the DNA only where a base has been modified. Hot NaOH (A>C) gives strong cleavage at A and weak at C. Additional base-specific reactions are available [3].

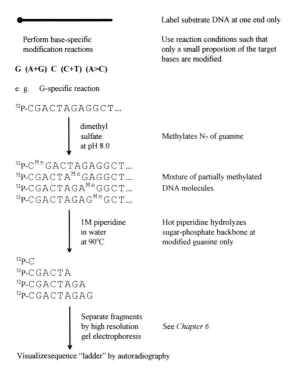

FIGURE 2.1: *Outline of chemical degradation method for DNA sequencing. Compare with chain termination method (*Figure 3.4*).*

The base-specific reactions are carefully designed to modify only a small proportion of the susceptible bases. Cleavage at the sites of modification then gives a set of end-labeled molecules ranging in length from one to several hundred nucleotides (*Figure 2.1*).

Compared with the chain termination method, the chemical degradation method has a number of advantages and disadvantages.

Advantages relative to chain termination:

- The sequence can be determined from within a DNA of unknown sequence, based only on a restriction map.
- The sequence can be determined very close to the labeled site (within 2–3 bases).
- The sequence is obtained from the original DNA molecule, rather than an enzymatic copy.

Disadvantages relative to chain termination:

* Less sequence is normally obtained.
* Reactions are generally slower and less reliable.
* Reactions require several hazardous chemicals.

For most purposes, these disadvantages outweigh the advantages. Consequently, the chemical degradation method is only used in specialist applications where it is important to sequence the original DNA molecule, rather than an enzymatic copy. This includes the study of DNA secondary structure, and the interaction of proteins with DNA. The chemical degradation reactions are sometimes performed on a DNA fragment of known sequence to produce a set of size markers, by comparison with which the size of other DNA molecules can be measured accurately.

References

1. Maxam, A.M. and Gilbert, W. (1977) A new method for sequencing DNA. *Proc. Natl Acad. Sci. USA,* **74**, 560–564.
2. Maxam, A.M. and Gilbert, W. (1980) Sequencing end-labelled DNA with base-specific cleavages. *Methods Enzymol.,* **65**, 499–560.
3. Ambrose, B.J.B and Pless, R.C. (1987) DNA sequencing: chemical methods. *Methods Enzymol.,* **152**, 522–538.

3 Chain Termination (Sanger Dideoxy) Method

3.1 Introduction

Chain termination (the Sanger or dideoxy method) is now by far the most widely used technique for sequencing DNA. There have been many minor modifications and improvements over the 20 years since first publication [1], but the principle remains the same. In the chain termination method, a DNA polymerase is used to extend a DNA strand (see *Figure 3.1*).

All the reagents needed for *in vitro* DNA synthesis are combined in the reaction, including a DNA polymerase and, additionally, a 2′, 3′-dideoxynucleotide is added (ddNTP, see *Figures 1.3* and *1.5*). ddNTPs can be incorporated by DNA polymerases into a growing DNA chain through their 5′ phosphate groups, just like dNTPs. However, they lack the 3′-OH group necessary for phosphodiester bond formation and chain elongation so the chain terminates at the precise point at which the ddNTP is incorporated (see *Figure 3.2*). Four sets of reactions are performed on each template, differing only in which of the four ddNTPs is added. The dNTP:ddNTP ratio is carefully selected so that the resulting labeled strands form a nested set of molecules up to several thousand bases long, each terminating at a specific base. These are separated according to size by high-resolution denaturing gel electrophoresis. This gives a 'ladder' of bands from which the DNA sequence can be read (see *Figure 6.1*).

This process is outlined in *Figure 3.3*. The key components of the reaction are listed in *Table 3.1*. Each of these contribute to the quality

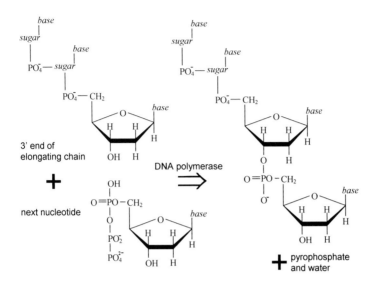

FIGURE 3.1: *Chain elongation. A DNA molecule is extended by the addition of a nucleotide to the 3′ carbon of the 3′-terminal sugar. This condensation reaction is catalyzed by DNA polymerase. A complementary template strand is also required (not shown).*

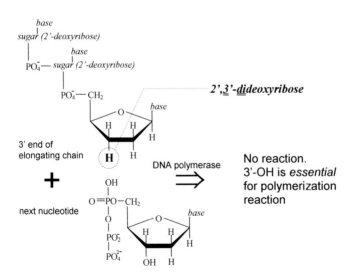

FIGURE 3.2: *Chain termination. The elongation reaction (Figure 3.1) requires a 3′-OH on the terminal sugar. Incorporation of a 2′, 3′-dideoxynucleotide, which lacks a 3′-OH (see Figure 1.3), therefore terminates chain elongation.*

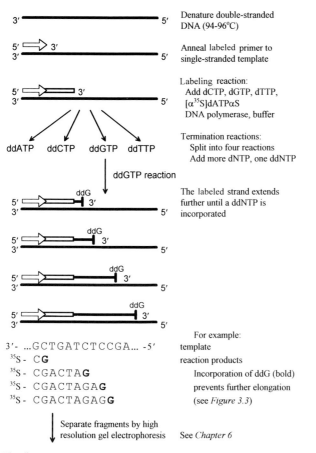

3' ━━━━━━━━━━━━━━ 5'	Denature double-stranded DNA (94-96°C)
5' ⇨ 3' 3' ━━━━━━━━━━━━━━ 5'	Anneal labeled primer to single-stranded template
5' ⇨━━ 3' 3' ━━━━━━━━━━━━━━ 5'	Labeling reaction: Add dCTP, dGTP, dTTP, [α³⁵S]dATPαS DNA polymerase, buffer
ddATP ddCTP ddGTP ddTTP	Termination reactions: Split into four reactions Add more dNTP, one ddNTP

ddGTP reaction

ddG 5' ⇨━━━┥ 3' 3' ━━━━━━━━━━━━━━ 5'	The labeled strand extends further until a ddNTP is incorporated
ddG 5' ⇨━━━━━┥ 3' 3' ━━━━━━━━━━━━━━ 5'	
ddG 5' ⇨━━━━━━━━┥ 3' 3' ━━━━━━━━━━━━━━ 5'	
ddG 5' ⇨━━━━━━━━━━━┥ 3' 3' ━━━━━━━━━━━━━━ 5'	

3'- ...GCTGATCTCCGA... -5'	For example: template
³⁵S - C**G**	reaction products
³⁵S - CGACTA**G**	Incorporation of ddG (bold)
³⁵S - CGACTAGA**G**	prevents further elongation
³⁵S - CGACTAGAG**G**	(see *Figure 3.3*)

Separate fragments by high resolution gel electrophoresis   See *Chapter 6*

Visualize sequence "ladder" by autoradiography

FIGURE 3.3: *Outline of chain termination method for DNA sequencing. Compare with cycle sequencing (Figure 3.4) and the chemical degradation method (Figure 2.1).*

TABLE 3.1: *The critical components of the chain termination method*

Key component	Discussed in
Primer	Section 4.2
DNA polymerase	Section 4.3
Label	Section 4.4
dNTPs and ddNTPs	Section 4.5 and 4.6
Template DNA	Chapter 5
Electrophoresis system	Chapter 6

PROTOCOL 3.1: Manual sequencing – the annealing reaction

Materials required
All versions:
- template DNA
- primer
- 5 × annealing buffer: 200mM Tris–HCI pH 7.5, 100 mM $MgCl_2$, 250 mM NaCl

Method B (double-stranded template)
- NaOH (e.g. 5 M)
- EDTA (e.g. 0.5 M)
- 3 M sodium acetate pH 4.5–5.5
- ethanol

Method C (double-stranded template)
- NaOH (fresh 1 M solution)
- *either* HCl (1 M) *or* TES buffer: 560 mM TES
  (2-[Tris(hydroxymethyl)methylamino]-1-ethanesulfonic acid), 240 mM HCl

Method
Use method A for single-stranded templates and either of methods B or C for double-stranded templates. Annealed templates can be stored overnight at –20°C or sequenced immediately (see *Protocol 3.2*).

A. Single-stranded template
1. Combine in a microcentrifuge tube:
   - single-stranded template DNA
     (see Section 5.1 and *Protocol 5*)   1 µg
   - primer   0.5–5 pmol
   - 5 × annealing buffer   2 µl
   - distilled water   to a final volume of 10 µl

2. Place tubes in a beaker containing approximately 1 l of water pre-heated to 65–70°C. Remove from the heat and allow to cool to below 30°C, over a period of 15–30 min. At this point, the annealing reaction is complete.

B. Double-stranded template
1. Denature DNA by adding to a microcentrifuge tube, in order:
   - double-stranded template DNA (~0.5–1 µg/kb length, e.g. 3–5 µg for typical plasmid) in 20–50 µl of TE (10 mM Tris–HC1 pH 8.0, 1 mM EDTA) or water
   - NaOH to 200 mM
   - EDTA to 0.2 mM

2. Neutralize and precipitate by adding, in order:
   - 3 M sodium acetate pH 4.5–5.5   0.1 vol.
   - ethanol   3 vol.

   Leave to precipitate at 0 to –90°C for at least 15 min. Centrifuge at 12 000 *g* in a microcentrifuge for 5 min. Remove and discard the supernatant. Rinse the pellet with 70% ethanol and air dry. It is not necessary to dry the pellet completely.

3. Resuspend and anneal the template by adding:
   - primer (e.g. 1 ml of 0.5–5 mM)   0.5–5 pmol
   - 5 × annealing buffer   2 µl
   - distilled water   to a final volume of 10 µl

   These reagents can be combined with each other and added together to the DNA pellet. Vortex briefly to resuspend the DNA and incubate at 37°C for 15–30 min.

C. Double-stranded template
This method requires the accurate neutralization of the NaOH. The exact amount of TES buffer or HCl required must be checked and the volume used in step 2 adjusted accordingly. Accurate pipeting is essential.

1. Denature the DNA by adding to a microcentrifuge tube, in order:
   - double-stranded template DNA (~0.5 –1 µg/kb length)   in 6 µl of TE or water
   - primer   1 µl of 0.5–5 µM
   - freshly prepared 1 M NaOH   1 µl

   Incubate at 68°C for 10 min.

2. Remove the tube from the water bath and immediately neutralize by adding 1–2 µl of TES buffer *or* 1 M HCl (see above). Incubate for 10 min at room temperature. Centrifuge briefly to collect condensation.

of the resulting sequence information, and are discussed individually in subsequent chapters.

A typical protocol for manual sequencing with ^{35}S is given in *Protocols 3.1* and *3.2*. In *Protocol 3.1*, two alternate methods are provided for double-stranded DNA templates [2–4]. Both methods denature the template with alkali. In method B, the template is then precipitated and resuspended in the presence of the primer. In method C, the alkali is neutralized by the addition of an acid, either HCl or TES. In *Protocol 3.2*, use either the dGTP mixes throughout or else the dITP mixes throughout (see Sections 4.6 and 8.3).

The keys to consistently successful sequencing are careful preparation and good reagents. Once the sequencing reactions are under way, the operator is under some time pressure. It is essential, therefore, that everything is to hand, pre-dispensed and stored at the correct temperature ready for use. I perform the reactions in batches of 10, using a microplate (Greiner Labortechnik cat. no. 653180). These small microplates have a well volume of about 15 μl, and a flat bottom giving good thermal contact with the heating block. Wetting the heating block further improves the contact. Use of a microplate makes pipeting much quicker. Do not use a microplate for the annealing reactions, as evaporative losses may reduce the volume.
For a batch of 10 annealed templates, I start the labeling reactions at 20–30 sec intervals. Dispensing the completed labeling reaction into the four termination mixes takes about 20–30 sec, as does adding the stop solution, so all the templates get approximately the same time at each stage.

3.2 Cycle sequencing

One of the limitations of the standard chain termination method is that only a single labeled DNA molecule is produced from each primer–template complex. The sensitivity of the method is limited, therefore, by the molar quantity of DNA template that can be used in the reaction. This is a major problem when sequencing large DNA templates (see Section 5.5) or purified DNA fragments, for example polymerase chain reaction (PCR) products (see Section 5.4). This limitation can be overcome by performing the sequencing reactions, denaturing the template DNA and repeating the reactions (refs 5–10, see *Figure 3.4*). In principle, this procedure can be repeated until one of the reaction components is exhausted. Repeated denaturation is performed most easily by thermal cycling – heating the reaction mix

---

PROTOCOL 3.2: Manual sequencing – the sequencing reactions

Materials required
- template DNA annealed to primer, from *Protocol 3.1*
- heating block, or water bath
- *either*   dNTP mix for dGTP sequencing: 1.5 μM dCTP, 1.5 μM dGTP, 1.5 μM dTTP
  or   dNTP mix for dITP sequencing: 1.5 μM dCTP, 3 μM dITP, 1.5 μM dTTP
- 100 mM dithiothreitol (DTT)
- [α-^{35}S]dATPαS (10–50 TBq mmol^{-1}, 370 MBq ml^{-1})
- Sequenase® or T7 DNA polymerase (7000 U ml^{-1})
- (optional) pyrophosphatase, sequencing grade (Amersham/USB)
- polymerase dilution buffer: 10 mM Tris–HCl pH 7.5, 5 mM DTT, 500 μg ml^{-1} bovine serum albumin (BSA)
- *either*   termination mixes for dGTP sequencing: 50 mM NaCl, 80 μM of each dNTP, + 8 μM of specific ddNTP
- *or*   termination mixes for dITP sequencing: 50 mM NaCl, 80 μM each of dATP, dCTP, dITP and dTTP + 8 μM of specific ddNTP, except ddG mix which is 50 mM NaCl, 160 μM dITP, 80 μM each of dATP, dCTP and dTTP + 1.6 μM ddGTP
- stop solution: 95% formamide, 20 mM EDTA pH 8.0, 0.05% bromophenol blue, 0.05% xylene cyanol FF

Method

A. *The labeling reaction*
1. Thaw the annealed template DNA (if frozen) and place on ice. Add the following:
   - template DNA from the annealing reaction   10 μl
   - dNTP mix   2 μl
   - 100 mM DTT   1 μl
   - [α-^{35}S]dATPαS   0.1–0.5 μl
   - DNA polymerase, diluted 10-fold in ice-cold dilution buffer   2 μl
   - (optional) pyrophosphatase   0.05 μl

The reaction starts when the polymerase is added, so this should be added last, when everything is ready for the termination reactions. Mix by pipeting, without introducing bubbles. Diluted polymerase is only stable for about 30 min. Pyrophosphatase, which is recommended when using dITP, can be pre-mixed with the polymerase.

2. Remove from ice and incubate at room temperature for 2–5 min. Performing the labeling reaction below room temperature, as the tube slowly warms, minimizes sequencing artifacts near the primer.

B. *The termination reactions*
1. Before starting the labeling reaction, prepare a microplate by labeling four sets of wells 'A', 'C', 'G' and 'T' and place on ice. Add 2.5 μl of the ddATP termination mix to the 'A' wells, 2.5 μl of the ddCTP mix to the 'C' wells, etc. Keep on ice.

2. Towards the end of the labeling reaction, pre-warm the microplate to 37–45°C for at least 1 min.

3. Immediately after the labeling reaction, transfer 3.5 μl of it into one each of the 'A', 'C', 'G' and 'T' wells. Mix by pipeting, without introducing bubbles. Incubate at 37–45°C for 3–5 min.

4. Add 4 μl of stop solution and store at –20°C until ready to load on the sequencing gel. Reactions are stable for a week.

---

to 94–96°C will denature the template; cooling below the melting temperature of the primer will allow it to anneal and the sequencing reaction to be repeated.

T7 DNA polymerase, the enzyme of choice for manual sequencing (see Section 4.3), is inactivated rapidly at 94°C, but the thermostable polymerases used for PCR are not. Use of a thermostable polymerase

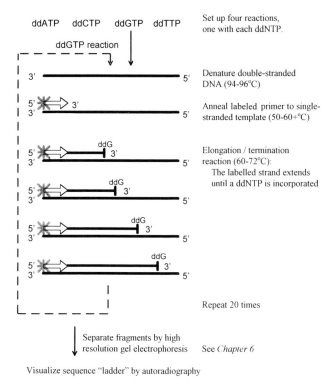

Set up four reactions, one with each ddNTP.

Denature double-stranded DNA (94-96°C)

Anneal labeled primer to single-stranded template (50-60+°C)

Elongation / termination reaction (60-72°C):
The labelled strand extends until a ddNTP is incorporated

Repeat 20 times

Separate fragments by high resolution gel electrophoresis   See *Chapter 6*

Visualize sequence "ladder" by autoradiography

FIGURE 3.4: *Outline of method for cycle sequencing. Compare with manual method (*Figure 3.3*) and chemical degradation method (*Figure 2.1*).*

and a thermal cycler ('PCR machine') allows the sequencing reactions to be repeated on the same template in an automated, cyclic fashion, requiring no additional manipulations by the researcher. In each cycle, the primer is annealed to the template, the normal dideoxy reactions are performed, then the newly synthesized double-stranded DNA is denatured by raising the temperature and the reaction is repeated. This allows a linear amplification of the amount of labeled material present in the reaction, and so less template DNA is required.

Cycle sequencing used to be much less consistent and reliable than conventional sequencing, but has improved dramatically in the last few years. In combination with a fluorescent label and semi-automated sequencers, this is the method used to generate most new sequence information. For example, the large-scale sequencing of the genome sequencing projects, expressed sequence tag (EST) and sequence tagged site (STS), generation has all been made possible by the substantial automation and parallel processing advantages of semi-automated sequencers.

Compared with conventional ('manual') dideoxy sequencing, cycle sequencing offers various advantages and disadvantages.

Advantages of cycle sequencing:

- An elevated reaction temperature, combined with routine use of dGTP analog (dITP or 7-deaza-GTP), reduces artifacts due to template secondary structure (see *Figures 3.5* and *4.5*).
- It requires much less template DNA, typically 50 fmol (the mass of 50 fmol of double-stranded DNA is ~ 33 ng/kb).
- If labeled primers are used, only sequences derived from the labeled primer will be detected. This method is therefore less sensitive to template impurities (but automated sequencers are very sensitive to such impurities, see Chapter 7).
- Reactions are easily automated.

Disadvantages of cycle sequencing:

- It requires the use of ^{32}P or ^{33}P, rather than ^{35}S, as neither T4 DNA kinase nor *Taq* polymerase will use thionucleotides efficiently. The relative merits of these isotopes are discussed in Section 4.4. Similarly, unmodified *Taq* polymerase incorporates fluorescently

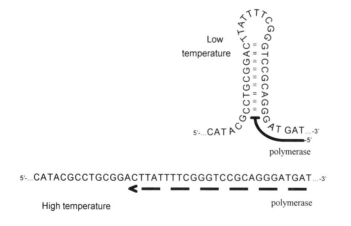

FIGURE 3.5: *Elevated temperature reduces sequence artefacts. Single stranded nucleic acids can form stem–loop structures by intramolecular base-pairing. This secondary structure can be destabilized by increasing the temperature. Secondary structure can affect both DNA replication, illustrated here, and also gel mobility.*

labeled dideoxynucleotides (DyeDeoxy™ terminators, Applied Biosystems) rather inefficiently, requiring their use at high concentrations. This is discussed further in Section 7.3.1.

- The thermostable polymerases available are less processive than T7 DNA polymerase and so read through homopolymeric runs such as oligo(T) less well.
- The quality of the sequence data obtained from cycle sequencing is generally lower than for conventional sequencing, though the computer enhancement of the data in automated sequencing more than compensates for this.
- Cycle sequencing is still generally less consistent and reliable than conventional sequencing.

For these reasons, cycle sequencing is only used in automated sequencing, or when the amount of template DNA is limiting. This includes the sequencing of PCR products (see Chapter 10) and also phage λ, cosmid and P1 phage clones, where the large size of the clone limits the molar concentration in the sequencing reaction [9]. Direct sequencing of PCR products is faster than subcloning and then sequencing. Furthermore, the average population of product molecules is sequenced, rather than a single one, which minimizes errors due to the low fidelity of *Taq* polymerase and also allows the direct detection of polymorphisms.

A typical protocol for cycle sequencing is given in *Protocols 3.3* and *3.4*. *Protocol 3.3* describes how to radiolabel an oligonucleotide primer with ^{32}P and *Protocol 3.4* uses this labeled primer for cycle sequencing. No further purification of the primer is required. *Protocol 3.4* uses *Taq* DNA polymerase, the most widely used thermostable polymerase. Alternative polymerases are available, but may incorporate nucleotide analogs with different efficiency, requiring modifications to the protocol [10, 11]. Addition of dimethylsulfoxide (DMSO) to a final concentration of 2–10% has been reported to improve the sequence from 'difficult' templates [12]. More detailed troubleshooting tips are presented in Chapter 8.

The program used for the thermal cycler depends on the model. Some instruments measure the temperature of the block, and so need extended times at each stage to allow the reaction to equilibrate to the block temperature, whereas others measure the sample temperature directly. The temperatures also depend on the thermostability and temperature optimum of the polymerase and the T_m of the primer (see Section 4.2.2). *Protocol 3.4* suggests 30 cycles, which is typical, but in principle this can be increased if the amount of template is limiting. Cycle numbers of 200 have been used for direct sequencing of genomic DNA [13], but then the thermal degradation of the polymerase in the

PROTOCOL 3.3: Radiolabeling the primer

Materials required
- 10 × T4 polynucleotide kinase (PNK) buffer: 500 mM Tris–HCl pH8.0, 100 mM MgCl$_2$, 100 mM DTT
- [γ-^{32}P]ATP approximately 185 TBq mmol^{-1}, 370 MBq ml^{-1} (e.g. Amersham PB10218)
- T4 polynucleotide kinase 5–10 U µl^{-1}

Method
1. Mix the following in a microcentrifuge tube, on ice:
   - 10 × PNK buffer   2 µl
   - primer   10 pmol
   - [γ-^{32}P]ATP (~ 185 TBq mmol^{-1}, 370 MBq ml^{-1})   10 pmol
   - distilled water   to 36 µl
   - T4 polynucleotide kinase (5–10 U µl^{-1})   0.5 µl
   - Take appropriate precautions for working with a radioisotope. This is 1.85 MBq (50 µCi) of [γ-^{32}P]ATP, which is 5 µl of Amersham PB10218.

2. Incubate at 37°C for 3 min.

3. Store at −20°C. Storage time is limited by radioactive decay.

These quantities make enough labeled primer to sequence 10 templates (see *Protocol 3.4*).

PROTOCOL 3.4: Cycle sequencing

Materials required
- thermal cycler and appropriate disposable tubes
- 100 mM DTT
- [α-^{35}S]dATPαS 10–50 TBq mmol^{-1}, 370 MBq ml^{-1}
- 10 × *Taq* reaction buffer: 300 mM Tris–HCl pH 9.5, 50 mM MgCl$_2$, 300 mM KCl, 0.5%(w/v) Tween-20 or Nonidet P-40 (NP-40)
- *Taq* DNA polymerase (5 U µl^{-1})
- four termination mixes: 50 mM NaCl, 80 µM each of dATP, dCTP, 7-deaza-GTP and dTTP + either 2 µM ddATP ('A' mix), 1 mM ddCTP ('C' mix), 0.2 mM ddGTP ('G' mix) or 2 mM ddTTP ('T' mix)
- mineral oil (not required with oil-free thermal cyclers)
- stop solution: 95% formamide, 20 mM EDTA pH 8.0, 0.05% bromophenol blue, 0.05% xylene cyanol FF

Method
1. Thaw the reagents and place them on ice. Label four sets of tubes (suitable for your thermal cycler) 'A', 'C', 'G' and 'T' and place on ice. Add 2 µl of the appropriate termination mix to each tube.

2. Mix the following on ice:
   - 10 × *Taq* reaction buffer   2 µl
   - template DNA   ~ 50 fmol
   - labeled primer from *Protocol 3.3*   1 pmol
   - distilled water   to 36 µl
   - *Taq* DNA polymerase (5 U µl^{-1})   0.5 µl

3. Add 8 µl of the mix from step 2 to each of the termination mixes prepared in step 1. Overlay with oil if required by your thermal cycler.

4. Program the thermal cycler. Parameters depend on the cycler and the application. A typical program might be 30 cycles:
   - 30 sec at 95°C   denaturing
   - 30 sec at 55°C   annealing
   - 60 sec at 72°C   extending/terminating
   the annealing temperature depends on the T_m of the primer (see Section 4.2).

5. Pre-heat the thermal cycler to 95°C, add the tubes and run the program. Remove the tubes from the thermal cycler. Optionally, separate the aqueous phase from oil by chloroform extraction.

6. Add 5 µl of stop solution and store at −20°C until ready to load on the sequencing gel. Reactions are stable for a few days.

denaturation step becomes significant. It is good practice to keep the denaturation step as short as possible in any case; the exact time and temperature depend, in part, on the model of thermal cycler you use.

References

1. Sanger, F., Nicklen, S. and Coulson, A.R. (1977) DNA sequencing with chain-terminating inhibitors. *Proc. Natl Acad. Sci. USA*, **74**, 5463–5467.
2. Chen, E.Y. and Seeburg, P.H. (1985) Supercoil sequencing: a fast and simple method for sequencing plasmid DNA. *DNA*, **4**, 165–170.
3. Zimmermann, J., Voss H., Schwager, C., Stegemann, J., Erfle, H., Stucky, K., Kristensen, T. and Ansorge, W. (1990) A simplified protocol for fast plasmid DNA sequencing. *Nucleic Acids Res.*, **18**, 1067.
4. Rouer, E. (1994) Direct neutralization of alkaline-denatured plasmid DNA in sequencing protocol by the sequencing reagent itself. *Nucleic Acids Res.*, **22**, 4844.
5. Innis, M.A., Myambo, K.B., Gelfand, D.H. and Brow, M.A.D. (1988) DNA sequencing with *Thermus aquaticus* DNA polymerase and direct sequencing of polymerase chain reaction amplified DNA. *Proc. Natl Acad. Sci. USA*, **85**, 9436–9440.
6. Murray, V. (1989) Improved double-stranded DNA sequencing using the linear polymerase chain reaction. *Nucleic Acids Res.*, **17**, 88–89.
7. Levedakou, E.N., Landegren, U. and Hood, L.E. (1989) A strategy to study gene polymorphism by direct sequence analysis of cosmid clones and amplified genomic DNA. *BioTechniques*, **7**, 438–442.
8. Carothers, A.M., Urlaub, G., Mucha, J., Grunberger, D. and Chasin, L.A. (1989) Point mutation analysis in a mammalian gene: rapid preparation of total RNA, PCR amplification of cDNA, and Taq sequencing by a novel method. *BioTechniques*, **7**, 494–499.
9. Voss, H., Zimmerman, J., Schwager, C., Erfle, H., Stegemann, J., Stucky, K. and Ansorge, W. (1990) Automated fluorescent sequencing of lambda DNA. *Nucleic Acids Res.*, **18**, 5314.
10. Mead, D.A., McClary, J.A., Luckey, J.A., Kostichka, A.J., Witney, F.R. and Smith, L.M. (1991) *Bst* DNA polymerase permits rapid sequence analysis from nanogram amounts of template. *BioTechniques*, **11**, 76–87.
11. Sears, L.E, Moran, L.S., Kissinger, C., Creasey, T., Perry-O'Keefe, H., Roskey, M., Sutherland, E. and Slatko, B.E. (1992) CircumVent thermal cycle sequencing and alternative manual and automated DNA sequencing protocols using the highly thermostable VentR (exo–) DNA polymerase. *BioTechniques*, **13**, 626–633.
12. Winship, P.R. (1989) An improved method for directly sequencing PCR amplified material using dimethyl sulphoxide. *Nucleic Acids Res.*, **17**, 1266.
13. Cairns, M.J. and Murray, V. (1994) Dideoxy genomic sequencing of a single-copy mammalian gene using more than two hundred cycles of linear amplification. *BioTechniques*, **17**, 910–914.

4 Instrumentation and Reagents

4.1 Getting started – sequencing kits

While it is possible to obtain each reagent separately, by far the simplest and quickest way to begin sequencing is to buy a good sequencing kit and follow the instructions. Numerous kits are available from a variety of manufacturers, for 'conventional' manual sequencing using [α-^{35}S]dATPαS as the radiolabel, and also for cycle sequencing and a variety of nonradioactive methods.

For manual sequencing, I can highly recommend Amersham/USB's Sequenase® v2.0 kit. This kit gives extremely good, consistent results and comes with an excellent protocol manual and troubleshooting guide, thoughtfully printed on waterproof, tear-resistant material. This kit is based on Amersham/USB's Sequenase® v2.0 enzyme, a T7 DNA polymerase genetically modified to eliminate the 3'–5' exonuclease activity of the unmodified enzyme [1]. For cycle sequencing with an ABI semi-automated sequencer, I use ABI's 'PRISM' dye terminator kits with Amplitaq® FS. Sequencing using semi-automated sequencers is discussed in Chapter 7.

In addition to a sequencing kit, other equipment and reagents are needed, as listed in *Table 4.1*.

Purification of template DNA, protocols for the reactions and gel electrophoresis are covered in other chapters. The other key reagents are discussed below.

TABLE 4.1: Equipment and reagents for manual and cycle sequencing

Manual	Cycle
Equipment	*Equipment*
Micropipettes	Micropipettes
Microcentrifuge tubes	Microcentrifuge tubes
Heating block or water bath	Thermal cycler
Microplates (optional)	
Polyacrylamide gel electrophoresis system and reagents	
Reagents	*Reagents*
Template DNA	Template DNA
Oligonucleotide primers	Oligonucleotide primers
Annealing buffer	*Taq* reaction buffer
Dithiothreitol (DTT)	Dithiothreitol (DTT)
dNTP mix	
Termination mixes	Termination mixes
DNA polymerase, e.g. T7	DNA polymerase, e.g. *Taq*
Stop solution	Stop solution
Labeled nucleotide	Label, e.g. on primer
e.g. [α-^{35}S]dATPαS	
Pyrophosphatase (optional)	

4.2 Oligonucleotide primers

All DNA polymerases used for sequencing require a primer. This is a short, synthetic, single-stranded DNA molecule of known sequence, typically 18–24 bases long, which anneals to the complementary sequence on the template DNA. 'Universal' primers are included in most sequencing kits. These are typically the M13–20 primer 5'-GTAAAACGACGGCCAGT-3' and the M13 reverse primer 5'-AACAGCTATGACCATG-3'. These primers are 'universal' in that their complementary sequences have been engineered into most common cloning vectors in such a way that the sequence of both ends of an unknown segment of DNA cloned into the polylinker can be determined by using these two primers (see *Figure 4.1*).

The cost of oligonucleotide synthesis has fallen dramatically, and is still falling, so that synthesis of custom primers to complete a sequencing project is routine. It is now no longer prohibitively expensive to sequence a DNA molecule by sequencing the ends, designing primers complementary to the ends of this newly determined sequence, sequencing with these new primers and so sequencing in from the ends in a stepwise fashion. This strategy, known as 'primer walking', is discussed in Section 11.3.

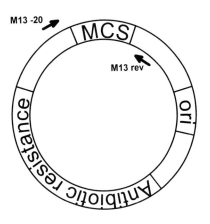

FIGURE 4.1: *Plasmid vector with multiple cloning site flanked by 'universal' primer sequences. The key features of a plasmid vector are an origin of replication (ori), a selectable marker, e.g ampicillin resistance, and a set of convenient restriction sites for cloning (multiple cloning site, MCS). Flanking the MCS with the 'universal' primer sequences allows both ends of any cloned insert to be sequenced using these standard primers.*

4.2.1 Primer design

The primer must anneal to exactly one site on the template DNA. In the standard salt and temperature conditions of the sequencing reactions, the stability of the primer–template complex depends on the length and sequence of the primer. For sequencing, a minimum length of 18 bases is normal, with a base composition at least 50% G+C. Longer or more GC-rich primers will form more stable hybrids. A simple formula for calculating the melting temperature (T_m) in °C of an oligonucleotide is

$$T_m = 2(A+T) + 4(G+C) \quad [2]$$

A more accurate formula is

$$T_m = 81.5 + 16.6(\log_{10}[Na^+]) + 0.41(\%G+C) - 600/L$$

where $[Na^+]$ is the molar concentration of monovalent cations such as sodium ions and L is the length of the oligonucleotide in bases [3].

In practice, the composition of the sequencing reaction is more complex than this, and both formulae should be regarded as approximations only. A typical annealing temperature might be 3–5°C below the annealing temperature calculated from the first formula or 0–10°C higher than the annealing temperature calculated from the second formula. This will often give satisfactory results first time, but can be optimized by trying alternative annealing temperatures a few degrees above and below the calculated value. Too high an annealing temperature will give a weak or no sequence, too low will give high background.

It is, of course, vital that the primer is efficiently synthesized, so that essentially all the primer molecules are of the same length and sequence. Inefficient synthesis can cause problems. For example, it can result in significant contamination with primers one base shorter than the nominal length. These will give a set of 'ghost' bands identical to the real bands, but with every band one base shorter, which may make the sequence unreadable (*Figure 4.2*). Similarly, if

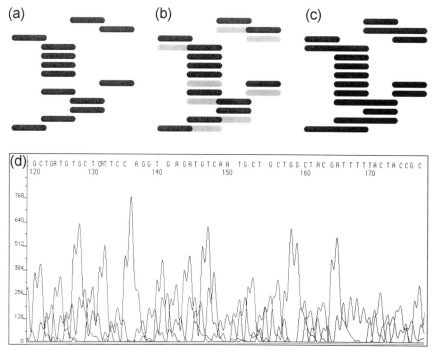

FIGURE 4.2: Mixed sequence. Primers of nonuniform length, mixed template and templates with two primer-binding sites will all give a mixed sequence. For example, primer molecules one base shorter at the 5′ end than the nominal length will give a sequence ladder with each band shifted by one base. **(a)** Ideal sequence; all primer molecules are the same length. **(b)** Sequence with a small proportion of primer molecules one base short. **(c)** Sequence with half the primer molecules one base short. **(d)** The same effect on automated sequencing.

the template DNA contains a second binding site for the primer, this will give two superimposed sets of sequence – one derived from each primer-binding site. Inefficient synthesis is a common problem only with long primers (30+ bases). Such oligonucleotides should therefore be purified before use as sequencing primers, for example by high-performance liquid chromatography (HPLC).

4.2.2 Primer design for cycle sequencing

Cycle sequencing works on the same general principles as PCR (Chapter 10), though the template DNA is amplified linearly rather than exponentially. Primer design for cycle sequencing therefore follows the same rules as primer design for PCR. Computer programs are available to assist with primer design, but all that is really needed is to follow some simple guidelines. Primers should typically be 20–25 nucleotides in length with 50–60% G+C content, which allows a reasonably high annealing temperature to be used. It is helpful if all sequencing primers are designed to have a similar annealing temperature, as then every reaction can use the same program in the thermal cycler. The end two bases at the 3′ end should preferably be G or C. The primer sequence must not be self-complementary, especially at the 3′ end.

4.3 DNA polymerase

The most commonly used polymerases for sequencing are T7 or modified T7 DNA polymerase (Sequenase® v2.0) for manual sequencing and *Thermus aquaticus* (*Taq*) DNA polymerase for cycle sequencing [1, 4, 5].

Thermostable polymerases such as *Taq* have the advantage that the reactions can be run at high temperature, minimizing artifacts due to secondary structure. Secondary structure occurs where the template DNA can anneal to itself, resulting in a region of double-stranded DNA which must be separated by the DNA polymerase as it passes through this sequence. The double-stranded region therefore forms a barrier to the progress of the polymerase. If the polymerase stops here, this will result in a termination that is *not* due to the incorporation of a specific ddNTP and so will give poor sequence, typically appearing as a band in all four lanes (see Section 8.1). The stability of double-stranded regions is temperature dependent, being destabilized at higher temperatures, so performing the sequencing reactions at high temperature minimizes these secondary structure artifacts (see *Figure 3.5*). A number of thermostable polymerases are available, with slightly different characteristics such as fidelity, half-life at high temperature and utilization of base analogs [6–8].

For manual sequencing, the enzyme of choice is Sequenase® v2.0 (Amersham/USB), a modified T7 DNA polymerase from which the 3′–5′ exonuclease activity has been removed by deleting amino acids 118–145 [1]. In practice, though, I find little difference between Sequenase and unmodified T7 polymerase (Pharmacia). These enzymes both show high speed, good processivity and will readily incorporate base analogs such as ddNTPs, dITP and 7-deaza-dGTP, giving uniform band intensities and minimizing background bands.

4.4 Label

The newly synthesized strands must be labeled in some way so that they can be detected after gel electrophoresis. This label can be incorporated at the 5′ end, at the 3′ end or in the middle (*Figure 4.3*). For traditional manual sequencing, a radioisotope is incorporated in the middle as the chain is synthesized, by including a labeled deoxynucleotide triphosphate in the reaction mix. The sequencing reaction then comprises three parts: the annealing reaction in which the primer hybridizes to the template; the labeling reaction in which the primer is extended a short distance using limiting concentrations of dNTPs together with a single radiolabeled dNTP; and the

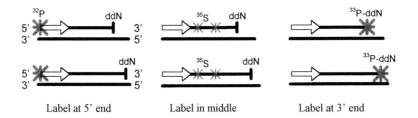

Label at 5' end   Label in middle   Label at 3' end

FIGURE 4.3: *Position of label. The newly synthesized strand can be labeled at the 5′ end by labeling the primer, in the middle by incorporating labeled dNTP analogs, or at the 3′ end by using labeled ddNTP analogs. An example of a typical radioisotope used at each position is shown. Fluorescent labeling is also possible at each of these positions (see Chapter 7).*

extension/termination reaction in which the 'extended primers' are extended further in the presence of both dNTPs and ddNTPs, leading to sequence-specific chain terminations. The principal advantage of this method is that multiple radiolabels are incorporated into each chain, allowing the use of relatively low-energy beta (β) particle-emitting isotopes such as ³⁵S.

The isotope used affects both the amount of sequence information that can be read from a single reaction and the exposure time required. The most widely used isotope is ³⁵S, in the form of [α-³⁵S]dATPαS (Amersham, Du Pont/NEN), whose relatively low energy β emission gives sharp bands on the autoradiogram, allowing longer and more accurate reading of the sequence [9].

End-labeling methods give a lower specific activity for the synthesized strand, so they must use more active isotopes such as ³²P. Unfortunately, the β emission from ³²P is much more penetrating, giving more diffuse bands, and so less sequence information can be obtained from a given reaction (see *Figure 4.4*). A more recently available, more expensive isotope, ³³P, has an intermediate emission energy and so gives somewhat better resolution [10, 11]. [γ-³³P]ATP is available for 5' labeling oligonucleotides, using T4 polynucleotide kinase, and [α-³³P]ddNTPs are now available for 3' labeling. Either of these labels allows the incorporation of only a single label into each

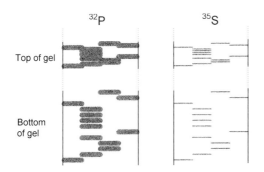

FIGURE 4.4: *Emission energy and resolution. At the top of a sequencing gel, further from the primer, bands are closely spaced. Diffuse bands from high-energy isotopes such as ³²P are not well resolved, particularly for short blocks of the same base. With the same band spacing, the sharper bands from ³⁵S are still well resolved. At the bottom of the gel, closer to the primer, bands are spaced more widely and both sequence ladders can be read.*

newly synthesized strand, so these methods are more suitable for cycle sequencing, where multiple labeled molecules are synthesized from each template strand.

Use of radioisotopes has two major disadvantages – safety and limited shelf life. Use of radioisotopes clearly poses a contamination hazard, and all tips and solutions must be handled and disposed of accordingly. In particular ^{35}S-labeled thionucleotides can break down into volatile radioactive by-products. The rules governing usage of radioisotopes vary, but in some areas are so severe as to provide a strong impetus to move to nonradioactive methods, even when such methods themselves perhaps require more expensive or hazardous reagents. Radioisotopes used for sequencing have limited half-lives (*Table 4.2*). The stock of radiolabel must, therefore, be replenished continuously, which can lead to additional cost and exacerbate waste-disposal problems. Radiolabeled sequencing reactions themselves must be run on a gel promptly.

Nonradioactive labels are also available. These are either haptens, which are then detected by histochemical or chemiluminescent methods, or fluorescent dyes, which can be visualized directly. Haptens are incorporated just like a radioisotope; at the 5′ end by using labeled primers or in the middle by using a dNTP analog labeled with a hapten. Suitable haptens include biotin or digoxigenin.

The recent introduction of fluorescent labels, and the technology to detect them in real time during electrophoresis, was a major advance, allowing much higher throughput and automation. Fluorescent labels can be incorporated by using labeled primers, a labeled dNTP or labeled ddNTPs (DyeDeoxys™, Perkin-Elmer/ABI) [12–15]. Many different fluorochromes are available, so it is possible to label the four termination reactions differentially (ddA, ddG, ddC and ddT), then combine them and run them in a single lane on the sequencing gel. If fluorescently labeled ddNTPs are used, the four termination reactions can be performed in the same tube, as a single mixed reaction. If labeled primers or dNTPs are used, then the four reactions are

TABLE 4.2: *Emission energies and half-lives of radioisotopes used in DNA sequencing*

	Max. emission energy (MeV)	Half-life (days)
^{32}P	1.709	14.3
^{33}P	0.249	25.4
^{35}S	0.167	87.1

performed separately, but may be combined before electrophoresis. Use of fluorescent labels requires an appropriate detection device (sequencing machine, see Chapter 7), which demands a high capital outlay.

There is an additional major benefit to using labeled ddNTPs. This is that only those reaction products which have been terminated by incorporation of a ddNTP will be labeled, and so only these will be detected. Spurious bands from false terminations will not be detected. This eliminates one major class of sequencing artifacts. Furthermore, another major class of artifacts, compressions, can also be avoided by the use of base analogs such as dITP. Inclusion of these analogs in the sequencing reactions eliminates compressions by destabilizing secondary structure in the DNA that leads to anomalous gel mobility, but dITP increases the number of false terminations. However, with labeled ddNTPs, these false terminations are not detected, so base analogs can be used routinely. Sequence artifacts such as false terminations and compressions are discussed further in Chapter 8, and the relative merits of different fluorescent sequencing chemistries are discussed in Chapter 7.

Finally, sequencing reactions can be performed with no label at all, the unlabeled reaction products being detected by silver staining. This is the basis of Promega's Silver Sequence™ kit. Direct detection is less sensitive than radiolabeling, requiring about 10 times as much product. This is not normally a limitation for standard templates.

The advantages and disadvantages of the various labeling methods are summarized in *Table 4.3*.

4.5 dNTPs and ddNTPs

The quality of the dNTPs and ddNTPs used, and their precise concentrations and molar ratios in the reaction mixes are absolutely critical for obtaining good sequence information. I therefore recommend purchasing a reputable kit, such as Amersham's Sequenase® v2.0 kit, in which all these reagents are pre-mixed and ready to use. As these mixes become exhausted, you may wish to replace them with 'home-made' mixes of the individual components. This approach can yield considerable cost savings, but the time and effort involved in optimizing the conditions may largely negate this. For this reason, I use the Sequenase kit for all my manual sequencing, rather than 'home-made' mixes. I find that the Sequenase

TABLE 4.3: Summary of labeling and detection strategies

Label site		General	Label type		
			Radioisotope	Hapten	Fluorochrome
Primer	Advantages	Uniform bands	Diffuse bands with ^{32}P	Labeled primers can be stored Rapid results	Labeled primers can be stored Uniform peak heights (ABI)
	Disadvantages	Only one label/molecule	Difficult with ^{35}S Radiation hazard	Requires technically demanding processing	Cost of primers Needs four reactions
dNTP	Advantages	Sensitive as multiple labels per molecule	Allows use of ^{35}S, giving sharp bands ^{35}S may require long exposure times	Rapid results	All reaction products exactly comparable Needs four reactions and four lanes (LI-COR)
	Disadvantages	Not suitable for cycle sequencing Faint close to primer	Radiation hazard	Requires technically demanding processing	
ddNTP	Advantages	Co-terminations not detected			Needs only one reaction/sample (ABI)
	Disadvantages	Only one label/molecule	Radiation hazard	Not available	Uneven peak heights (ABI)
No label	Advantages	Requires no label!			
	Disadvantages	Limited sensitivity Difficult to automate			

enzyme itself is always the first kit component to be used up, so we buy more T7 DNA polymerase (Pharmacia) and continue to use the reaction mixes from the original kit.

For 'home-made' nucleotide mixes, note that the optimum concentration of dNTPs and ddNTPs depends on the DNA polymerase used. Some polymerases discriminate against ddNTPs, preferentially incorporating dNTPs, so these require higher concentrations of ddNTPs to give chain termination at the same average chain length. Similarly, some polymerases incorporate dITP or 7-deaza-dGTP less efficiently than standard dNTPs, and so need higher concentrations of these nucleotide analogs.

4.6 dITP and 7-deaza-dGTP

Single-stranded DNA molecules tend to anneal to complementary sequences, located either on other DNA molecules or elsewhere on the same molecule. This is the basis of the annealing reaction, in which the primer anneals to a complementary region of the template. The products of the sequencing reaction are separated electrophoretically on an acrylamide gels and it is essential here that they remain single-stranded. Formation of hairpin loops by intramolecular hybridization may substantially alter the electrophoretic mobility of the molecule, and so affect the quality of the derived sequence information. This leads to a 'compression', in which band spacing is altered (see Section 8.3). G–C hybrids are significantly more stable, so compressions are found in regions of high G+C content.

This artifact can be reduced by using base analogs which form less stable hybrids than the standard nucleotide bases. Either dITP or 7-deaza-dGTP can be used in place of dGTP, destabilizing G–C base pairs and so reducing or eliminating compressions [16, 17] (see *Figure 4.5*). Some laboratories routinely use these analogs for all their sequencing, others only to re-sequence a region which has been found to produce compressions. Using Amersham's Sequenase® v2.0 kit, I find that dITP reactions cannot be read as far as dGTP ones, and are more liable to co-terminations (Section 8.2). I therefore use dGTP routinely, and re-sequence suspicious regions using dITP. I use ABI's AmpliTaq® FS kit for cycle sequencing. This kit contains dITP as standard. An alternative method for resolving compressions is the use of formamide gels (Section 6.5.3).

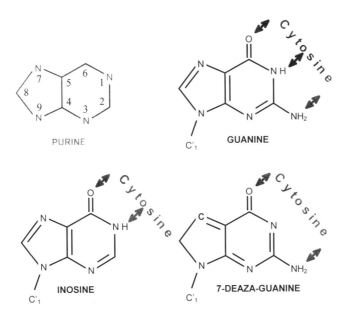

FIGURE 4.5: *The structure of base analogs used to resolve compressions. Inosine and 7-deaza-guanine form only two hydrogen bonds when base pairing with cytosine. These base pairs are therefore less stable than G–C base pairs, which form three hydrogen bonds. Secondary structure is therefore less stable when dITP or 7-deaza-dGTP is used in place of dGTP in the sequencing reactions. This minimizes compressions.*

4.7 Pyrophosphatase

The products of the sequencing reactions are liable to pyrophosphorolysis, which is degradation by reaction with pyrophosphate. This can occur in a sequence-specific fashion, leading to the loss of specific bands. This is only a serious problem with dITP sequencing and can be overcome by the addition of pyrophosphatase to the sequencing reaction to destroy the pyrophosphate [18]. Both the Amersham Sequenase® v2.0 and the ABI AmpliTaq® FS kits include pyrophosphatase; it can also be obtained separately.

References

1. Tabor, S. and Richardson, C.C. (1989) Selective inactivation of the exonuclease activity of bacteriophage T7 DNA polymerase by *in vitro* mutagenesis. *J. Biol. Chem.*, **264**, 6447–6458.

2. Furrer, B., Candrian, U., Wieland, P. and Lüthy, J. (1990) Improving PCR efficiency. *Nature*, **346**, 324.

3. Zhu, Y.S., Isaacs, S.T., Cimino, G.D. and Hearst, J.E. (1991) The use of exonuclease III for polymerase chain reaction sterilization. *Nucleic Acids Res.*, **19**, 2511.

4. Tabor, S. and Richardson, C.C. (1987) DNA sequence analysis with a modified bacteriophage T7 DNA polymerase. *Proc. Natl Acad. Sci. USA*, **84**, 4767–4771.

5. Innis, M.A., Myambo, K.B., Gelfand D.H. and Brow, M.A.D. (1988) DNA sequencing with *Thermus aquaticus* DNA polymerase and direct sequencing of polymerase chain reaction amplified DNA. *Proc. Natl Acad. Sci. USA*, **85**, 9436–9440.

6. Sears, L.E., Moran,, L.S., Kissinger, C., Creasey, T., Perry-O'Keefe, H., Roskey, M., Sutherland, E. and Slatko, B.E. (1992) CircumVent thermal cycle sequencing and alternative manual and automated DNA sequencing protocols using the highly thermostable VentR (exo–) DNA polymerase. *BioTechniques*, **13**, 626–633.

7. Mead, D.A., McClary, J.A., Luckey, J.A., Kostichka, A.J., Witney, F.R. and Smith, L.M. (1991) *Bst* DNA polymerase permits rapid sequence analysis from nanogram amounts of template. *BioTechniques*, **11**, 76–87.

8. Tabor, S. and Richardson, C.C. (1995) A single residue in DNA polymerases of the *E. coli* DNA polymerase I family is critical for distinguishing between deoxy- and dideoxyribonucleotides. *Proc. Natl Acad. Sci. USA*, **92**, 6339–6343.

9. Biggin, M.D., Gibson, T.J. and Hong, G.F. (1983) Buffer gradient gels and ^{35}S label as an aid to rapid DNA sequence determination. *Proc. Natl Acad. Sci. USA*, **80**, 3963–3965.

10. Zagursky, R.J., Conway, P.S. and Kashdan, M.A. (1991) The use of ^{33}P for Sanger DNA sequencing. *BioTechniques*, **11**, 36–38.

11. Evans, M.R and Read, C.A. (1992) ^{32}P, ^{33}P and [35]S: selecting a label for nucleic acid analysis. *Nature*, **358**, 520–521.

12. Smith, L.M., Sanders, J.Z., Kaiser, R.J., Hughes, P., Dodd, C., Connell, C.R., Heiner, C., Kent, S.B. and Hood, L.E. (1986) Fluorescence detection in automated DNA sequence analysis. *Nature*, **321**, 674–679.

13. Middendorf, L.R., Bruce, J.C., Bruce, R.C., *et. al.* (1992) Continuous, on-line DNA sequencing using a versatile infrared laser scanner/electrophoresis apparatus. *Electrophoresis*, **13**, 487–494.

14. Prober, J.M., Trainor, G.L., Dam, R.J., Hobbs, F.W., Robertson, C.W., Zagursky, R.J., Cocuzza, A.J., Jensen, M.A. and Baumeister, K. (1987) A system for rapid DNA sequencing with fluorescent chain-terminating dideoxynucleotides. *Science*, **238**, 336–341.

15. Lee, L.G., Connell, C.R., Woo, S.L., Cheng, R.D., McArdle, B.F., Fuller, C.W., Halloran, N.D. and Wilson, R.K. (1992) DNA sequencing with dye-labeled terminators and T7 DNA polymerase: effect of dyes and dNTPs on incorporation of dye-terminators and probability analysis of termination fragments. *Nucleic Acids Res.*, **20**, 2471–2483.

16. Barr, P.J., Thayer, R.M., Laybourn, P., Najarian, R.C., Seela, F. and Tolan, D.R. (1986) 7-deaza-2'-deoxyguanosine-5'-triphosphate: enhanced resolution in M13 dideoxy sequencing. *BioTechniques*, **4**, 428.

17. Mizusawa, S., Nishimura, S. and Seela, F. (1986) Improvement of the dideoxy chain termination method of DNA sequencing by use of deoxy-7-deazaguanosine triphosphate in place of dGTP. *Nucleic Acids Res.*, **14**, 1319–1324.

18. Tabor, S. and Richardson, C.C. (1990) DNA sequence analysis with a modified bacteriophage T7 DNA polymerase. Effect of pyrophosphorolysis and metal ions. *J. Biol. Chem.*, **265**, 8322–8328.

5 Template Preparation

5.1 Introduction

The first step of a sequencing reaction is to anneal the primer to the template DNA. This clearly requires the template strand not to be already annealed to its complement; in other words the template DNA must be single-stranded.

Purified DNA is double-stranded, unless it is purified from viruses which have single-stranded genomes, for example bacteriophage M13. DNA for sequencing must, therefore, either be purified in single-stranded form (using a vector with an appropriate origin of replication) or purified in double-stranded form and converted to single-stranded, most commonly by alkali or heat denaturation [1–4]. This is something of a compromise – double-stranded DNA is more convenient to purify, can be sequenced in either direction, and is more useful for other purposes, but the quality of the sequence information obtained is generally somewhat lower than for single-stranded DNA, and less can be read from each reaction.

5.2 Preparing single-stranded DNA templates

Conceptually, the simplest way to obtain single-stranded DNA templates is to subclone the fragment of interest into an M13 vector, purify the phage particles and extract the single-stranded DNA. However, M13 vectors have disadvantages: they are liable to rearrange or delete large inserts and it can be difficult to purify the double-stranded 'replicative form' version required for restriction enzyme digestion and subcloning.

Much more widely used today are 'phagemid' vectors. These are plasmids carrying two origins of replication, one from a plasmid and one from a single-stranded phage (see *Figure 5.1*). In the presence of a helper phage, these are expressed as single-stranded DNA and packaged into phage particles. Helper phage, such as M13 K07 [5], encode the proteins necessary for phage replication and growth, but themselves grow rather poorly, so that the helper phage DNA forms only a minor contaminant of the purified single-stranded DNA. Furthermore, the helper phage DNA does not include a region complementary to the 'universal' primers, and so the presence of small amounts does not adversely affect the sequencing reactions. *Protocol 5.1* describes how to grow single-stranded phage cultures from phagemid or M13 clones and *Protocol 5.2* describes how to purify these phage particles and how to purify single-stranded DNA from them.

Not all *Escherichia coli* strains can be infected by M13. Host strains must have an F or F' factor. Suitable strains include XL-1, DH5αF' and TG2. Strains such as DH5α and HB101, which do not have an F' cannot be used. It is important to select for the presence of the F', so as to ensure that all of the cells in the culture are capable of infection.

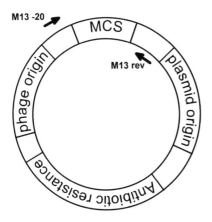

FIGURE 5.1: *Phagemid vector with multiple cloning site (MCS) flanked by 'universal' primer sequences. The key features of a phagemid vector are a plasmid origin of replication, a single-stranded phage origin of replication, a selectable marker, e.g. ampicillin resistance, and a set of convenient restriction sites for cloning (MCS). The phage origin will allow one strand to be purified following superinfection with a helper phage (see* Protocol *5.1). Note that only one of the two primers shown will be complementary to this strand, and so only one end can be sequenced using single-stranded template and a 'universal' primer.*

PROTOCOL 5.1: Bacteriophage culture for single-stranded templates

Materials required
- 37°C shaking incubator.
- sterile 10–15 ml tubes.
- LB liquid culture media: per liter: 10 g tryptone, 5 g yeast extract, 10 g NaCl, adjust pH to 7.0, autoclave.
- ampicillin stock solution: 100 mg ml⁻¹ in water, filter sterilized
- kanamycin stock solution: 100 mg ml⁻¹ in water, filter sterilized
- M13 K07 helper phage, > 10^{11} p.f.u. ml⁻¹ (e.g. Promega)

Method
Use method A for phagemids and method B for M13 clones.

A. Phagemids
1. Transform the phagemid DNA into an appropriate strain of *E. coli* (see text).

2. Inoculate 1 ml of LB containing 50–100 μg ml⁻¹ ampicillin with a single colony; grow overnight at 37°C with agitation.

3. Combine in a sterile 10–15 ml tube:
    - LB containing 50–100 μg ml⁻¹ ampicillin   2.5 ml
    - Bacterial culture containing phagemid   50 μl
    - M13 K07 helper phage   3 μl

4. Culture with agitation (e.g. 250–300 r.p.m.) at 37°C for 1 h.

5. Add kanamycin to 100 μg ml⁻¹.

6. Culture with agitation at 37°C for 6–8 h or overnight.

B. M13 clones
1. Grow 1 ml of culture of appropriate host cells (see text) to saturation (e.g. overnight).

2. Combine in a sterile 10–15 ml tube:
    - LB   2.5 ml
    - Bacterial culture   50 μl

3. Touch a sterile toothpick or wire loop to a single M13 plaque and use this to infect the culture.

4. Culture with agitation at 37°C for 5–7 h. Growth for extended periods may lead to the accumulation of deletion derivatives of the original M13 clone.

M13 infection is highly deleterious to the cells, so cells immune to infection can take over the culture rapidly. Many laboratory strains carry a selectable marker on the F′ for this reason, for example tet^R (XL-1). At the end of the procedure, the quality and quantity of single-stranded template DNA can be checked by agarose gel electrophoresis. Single-stranded DNA has a higher mobility on these gels than its double-stranded equivalent, and stains less well with ethidium bromide.

An alternative approach to obtaining single-stranded DNA is to purify double-stranded DNA and then to eliminate one strand selectively by enzymatic digestion. In an early step of M13 replication, the product of M13 gene II binds to the M13 origin and selectively nicks one strand (always the plus strand). The plus strand of double-stranded

PROTOCOL 5.2: Purifying single-stranded DNA from bacteriophage cultures

Materials required
- microcentrifuge
- microcentrifuge tubes
- 20% PEG 6000, 2.5 M NaCl
- TE buffer
- TE-saturated phenol
- chloroform ($CHCl_3$)
- freshly prepared 5 M ammonium acetate
- ethanol

Method

A. Purifying phage particles from the culture

1. Transfer 1.5 ml of the culture from *Protocol 5.1* to a microcentrifuge tube. Centrifuge at 12 000 *g* for 5 min. Transfer 1.3 ml of the supernatant to a fresh tube, avoiding the pellet.

2. Add 200 µl of 20% PEG 6000, 2.5 M NaCl to the supernatant and mix well. Allow to precipitate for 5–30 min at room temperature.

3. Centrifuge at 12 000 *g* for 5 min. Remove and discard the supernatant.

4. Centrifuge at 12 000 *g* for 30 sec. Carefully remove the remaining supernatant. Any remaining PEG may severely affect subsequent sequencing reactions.

5. Add 100 µl of TE and resuspend the pellet by vortexing. The phage particle suspension is stable indefinitely at 4°C.

B. Purifying DNA from phage particles

1. Phenol extract by adding an equal volume (100 µl) of TE-saturated phenol to the phage suspension from part A. Vortex for 30 secs. Centrifuge at 12 000 *g* for 5 min. Transfer the supernatant to a fresh tube, carefully avoiding the interface.

2. Chloroform extract by adding an equal volume (100 µl) of chloroform to the supernatant from step 1, vortex and centrifuge as above. Again, transfer the supernatant to a fresh tube, carefully avoiding the interface.

3. Ethanol precipitate by adding:
   - 5 M ammonium acetate   50 µl
   - ethanol   350 µl

4. Incubate on dry ice for 30 min or at –20°C overnight.

5. Centrifuge at 12 000 *g* for 10 min. Carefully remove and discard the supernatant. The pellet of single-stranded DNA may not be visible. Dry the pellet in air or under vacuum. It is not necessary to completely remove all of the ethanol.

6. Resuspend the DNA in 20–50 µl of TE or water. This DNA is stable indefinitely if stored at –20°C.

phagemid DNA can be nicked *in vitro* by purified M13 gpII. Treatment with an appropriate exonuclease then destroys the plus strand, leaving the purified minus strand (*Figure 5.2*). This is the basis of NBL's 'Quick-Strand' kit. Note that this method purifies the minus strand only, whereas use of a helper phage purifies the plus strand only. The two methods together therefore allow the purification of both single-stranded templates.

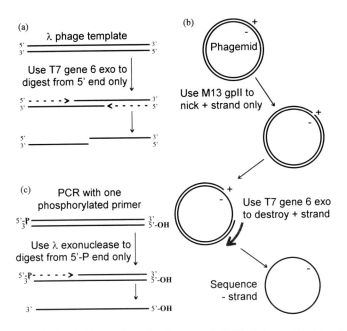

FIGURE 5.2: *Preparing single-stranded DNA from double-stranded DNA by enzymatic digestion. (a)T7 gene 6 exonuclease digests only the 5' end of double-stranded DNA. Phage λ is therefore degraded to two single-stranded molecules, each of approximately half the length of the original molecule. (b) In NBL's Quick-Strand kit, M13 gpII recognizes f1 and M13 replication origins and selectively nicks the + strand. Subsequent digestion with an appropriate exonuclease, e.g. T7 gene 6 or ExoIII degrades the + strand only, leaving the – strand. Note that single-strand rescue with a helper phage (Protocol 5.1) purifies the + strand only. (c) PCR using one 5' -phosphorylated and one unphosphorylated primer gives a product with one 5' end phosphorylated and the other unphosphorylated. λ exonuclease will digest from 5' -phosphorylated ends only, and so selectively hydrolyzes one strand.*

DNA purified from bacteriophage λ, which are linear double-stranded molecules, can be converted to single-stranded form by using bacteriophage T7 gene 6 exonuclease (Amersham/USB). This enzyme hydrolyzes double-stranded DNA from 5′ termini, resulting in half-length, single-stranded molecules (*Figure 5.2*). These can be used as single-stranded templates for sequencing any region of the molecule except the middle. This method can be applied to any double-stranded DNA, but λ DNA, typically 45–48 kb in length, is particularly difficult to sequence by conventional methods.

5.3 Preparing double-stranded DNA templates (plasmids)

Double-stranded templates must be of high purity. In particular, they must be free of short DNA and RNA fragments which could act as unwanted primers or templates. Plasmid DNA purified through a cesium chloride (CsCl) gradient is suitable, as are various proprietary methods, such as Qiagen preps (Qiagen) and Wizard preps (Promega). Small-scale preparations ('mini-preps'), for example by the alkaline lysis method, are not normally suitable without further purification using one of the above kits. One reasonably reliable protocol is given in ref. 6. Other purification methods such as the use of polyethyleneglycol (PEG) precipitation [7] seem to work well for some laboratories and not for others.

Double-stranded templates do not give good sequence as consistently as single-stranded templates and cannot normally be read as far. Co-termination ('pausing') can be a problem with all but the cleanest templates (see Section 8.2). Conversion of double-stranded DNA to single-stranded by enzymatic digestion is discussed in the preceding section.

5.4 PCR products

PCR products are linear double-stranded DNA molecules of modest size, typically less than 5 kb. As such they would seem to be acceptable templates. However, a PCR reaction mix contains significant quantities of reagents which must be completely removed from the product of interest before it can be sequenced successfully. These include dNTPs, primers and also reaction products other than the product of interest. Any PCR primers carried over from the PCR reaction may act as primers in the sequencing reaction, leading to a background of spurious bands. Carry-over of dNTPs will alter the dNTP:ddNTP ratios in the sequencing reactions. Reaction products other than the product of interest may act as unwanted templates. This is particularly a problem if the sequencing primer used corresponds to one of the PCR primers, as most reaction products will have the PCR primers at their ends even if their internal sequence is unrelated to the product of interest.

For these reasons, it is essential to purify the reaction product from the PCR reaction before using it as a sequence template. The minimum purification is a separation based on size, for example gel filtration or precipitation with ethanol or isopropanol, which will separate dNTPs and short oligonucleotides from the PCR product. This purification may be sufficient if the PCR product of interest is by far the major product (i.e. appears as a single band on an ethidium bromide-stained gel), and if the primers are short enough that they, and the primer dimers, are removed efficiently. I routinely gel-purify my PCR products before sequencing or subcloning. This not only removes low molecular weight contaminants, but also checks the size and amount of product synthesized, and separates it from any unwanted products. Many procedures and kits are available for purifying DNA from agarose. I use low gelling temperature agarose (BioGene), cut the desired band in a minimum volume of agarose, melt it at 70°C and the purify the DNA using the Wizard DNA Clean-up kit (Promega). In my hands, this DNA is at least as good a template as plasmid DNA.

A further problem applies to purified DNA fragments, such as PCR products, which is the amount of template DNA available. Conventional sequencing reactions require about 500–1000 fmol of template DNA (the mass of 50 fmol of double-stranded DNA is ~ 33 ng kb^{-1}). This quantity of DNA may be difficult to obtain. PCR products can be sequenced using ^{33}P cycle sequencing, which requires only about 50 fmol of template DNA. I use a little more template than this for my preferred method, which is fluorescent cycle sequencing – 10 ng/100 base length of the PCR product (~ 150 fmol), using Perkin-Elmer's dye terminator kit followed by analysis on an ABI 377 sequencer (see Section 7.3).

5.4.1 Single-stranded DNA templates from PCR products

Single-stranded DNA is a better template than double-stranded for DNA sequencing. PCR generates double-stranded DNA. Though this can be sequenced directly, the PCR procedure can be adapted to give single-stranded DNA for sequencing by using asymmetric PCR, or solid-phase capture or degradation by λ exonuclease.

Asymmetric PCR. PCR normally uses equimolar concentrations of each primer. If, instead, a huge excess of one primer is used relative to the other, then the final product will contain an excess of that strand, as single-stranded DNA [8] (see *Figure 5.3*). This asymmetric PCR can be performed directly on the original template, but

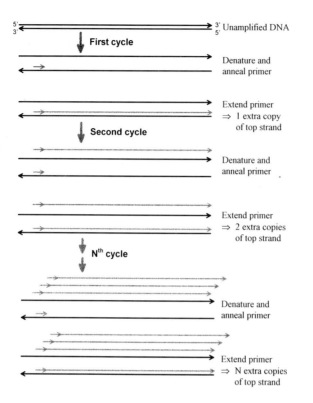

FIGURE 5.3: *Asymmetric PCR. PCR with only one primer gives linear amplification of one strand only (compare with* Figure 10.1*).*

reproducibly high yields of single-stranded DNA are hard to obtain. A more reliable approach is to purify the product of a normal PCR reaction and reamplify it for about 20 cycles with only one primer. Each cycle generates a single-stranded copy of the original PCR product, so 20 cycles gives a large excess of single-stranded copies of the original double-stranded PCR product.

Solid-phase capture. If one primer in a PCR reaction is labeled with an affinity tag such as biotin, it can be captured on suitably coated beads. Since only one strand is labeled, denaturing the DNA with NaOH and washing will leave only one strand attached to the beads. This single-stranded DNA can be eluted, or else sequenced directly while still attached to the beads. Conversely, the unlabeled strand can be recovered from the wash solution by ethanol precipitation [9–11]. Streptavidin-coated magnetic beads are available from Dynal [12]. Pre-incubation of the beads with a solution of BSA, casein or skimmed milk powder to block nonspecific binding may improve the sequence quality [13] (see *Figure 5.4*).

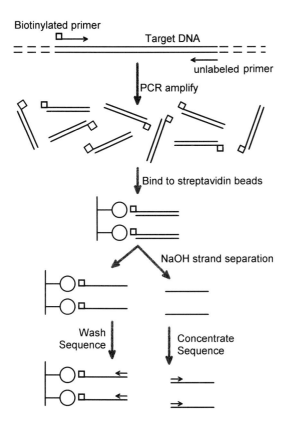

FIGURE 5.4: *Outline of template preparation by solid-phase capture. Tagging one primer with biotin allows both strands to be selectively purified.*

λ *exonuclease*. λ exonuclease digests double-stranded DNA from the 5′ end. The exonuclease requires or prefers a 5′ phosphate. Synthetic oligonucleotide primers have a 5′-OH, but can be phosphorylated by T4 polynucleotide kinase (e.g. *Protocol 3.3* with unlabeled ATP). PCR with one 5′-phosphorylated and one 5′-OH primer gives a product with one strand sensitive to λ exonuclease and the other resistant [14]. Digestion with the exonuclease therefore allows the selective purification of one strand (*Figure 5.2c*).

5.5 Large templates (lambda, cosmids, P1)

Lambda clones and cosmids are about 45–48 kb in size; P1 clones are 80–100 kb. Such large templates typically give very poor sequence, if any, by manual sequencing. The high molecular weight of these DNAs limits the molar quantity that can be used in a standard sequencing reaction. For example, 3 µg of 5 kb plasmid DNA is typically used in a 10–20 µl reaction. An equal molar quantity of a cosmid would be about 30 µg, which cannot be used for reasons of solubility, viscosity and effect on the sequencing gel.

Lambda DNA can be converted to a single-stranded template (see Section 5.2), but the most common solution is to subclone the region of interest into a phagemid vector and then sequence this smaller clone. Alternatively, large templates can be sequenced directly by cycle sequencing, in my hands yielding 300–500 bases of sequence (P1 templates of 80–100 kb) using dye terminators (ABI) and an ABI 377 sequencer. This is rather less sequence, and less consistent, than I obtain from plasmid templates.

5.6 Templates for semi-automated sequencing

As with manual sequencing, single-stranded templates are generally better than double-stranded. The major difference is that the quality of the DNA is absolutely critical for semi-automated sequencing, so that many templates that are perfectly adequate for manual sequencing will give poor results by semi-automated methods. Low levels of salt, ethanol and small nucleic acids seem to have particularly deleterious effects. I, therefore, always purify plasmid, cosmid and P1 templates for semi-automated sequencing using Qiagen columns (Qiagen). Qiagen also produce a comprehensive and extremely helpful manual covering various aspects of template purification for sequencing [14]. Promega have recently reformulated their Wizard DNA purification system to improve its performance for automated sequencing.

References

1. Wang, Y. (1988) Double-stranded DNA sequencing with T7 polymerase. *BioTechniques*, **6**, 843–845.
2. Casanova, J.-L., Pannetier, C., Javlin, C. and Kourilisky, P. (1990) Optimal conditions for directly sequencing double-stranded PCR products with Sequenase. *Nucleic Acids Res.*, **18**, 4028.

3. Kusukawa, N., Uemori, T., Asada, K. and Kato, I. (1990) Rapid and reliable protocol for direct sequencing of material amplified by the polymerase chain reaction. *BioTechniques*, **9**, 66–72.

4. Hsiao, K.-C. (1991) A fast and simple procedure for sequencing double-stranded DNA with Sequenase. *Nucleic Acids Res.*, **19**, 2787.

5. Vieira, J. and Messing, J. (1987) Production of single-stranded plasmid DNA. *Methods Enzymol.*, **153**, 3–11.

6. Kraft, R., Tardiff, J., Krauter, K.S. and Leinward, L.A. (1988) Using mini-prep plasmid DNA for sequencing double-stranded templates with Sequenase. *BioTechniques*, **6**, 544–546.

7. Sambrook, J., Fritsch, E.F. and Maniatis, T. (1989) Molecular Cloning: A Laboratory Manual. 2nd Edn. Cold Spring Harbor Laboratory Press, Cold Spring Harbor, NY.

8. Gyllensten, U.B. and Erlich, H.A. (1989) Ancient roots for polymorphism at the HLA-DQ alpha locus in primates. *Proc. Natl Acad. Sci. USA*, **86**, 9986–9990.

9. Mitchell, L.G. and Merril, C.R. (1989) Affinity generation of single-stranded DNA for dideoxy sequencing following the polymerase chain reaction. *Anal. Biochem.*, **178**, 239–242.

10. Hultman, T., Bergh, S., Moks, T. and Uhlen, M. (1991) Bidirectional solid-phase sequencing of in vitro-amplified plasmid DNA. *BioTechniques*, **10**, 84–93.

11. Hultman, T., Stahl, S., Hornes, E. and Uhlen, M. (1989) Direct solid phase sequencing of genomic and plasmid DNA using magnetic beads as solid support. *Nucleic Acids Res.*, **17**, 4937–4946.

12. Dynal, A.S. (1995) Biomagnetic Techniques in Molecular Biology. In *The Dynal Technical Handbook*. 2nd edn. Dynal AS, Oslo, Norway.

13. Lawson, V.A., McPhee, D.A. and Deacon, NJ. (1996). Elimination of sequence ambiguities by a single-step modification of a solid-phase, single-stranded sequencing protocol. *BioTechniques*, **21**, 356–358.

14. Higuchi, R.G. and Ochman, H. (1989) Production of single-stranded DNA templates by exonuclease digestion following the polymerase chain reaction. *Nucleic Acids Res.*, **17**, 5865.

15. Qiagen GmbH. (1995) The Qiagen Guide to Template Purification and DNA Sequencing. Qiagen, Hilden, Germany.

6 Gel Electrophoresis

6.1 Introduction

The DNA polymerases used in sequencing reactions can synthesize a nested set of labeled molecules extending more than 1000 bases from the primer, given a good template. This means that more than 1000 bases of sequence information are potentially available from a single sequencing reaction. In practice however, the subsequent gel electrophoresis is unable to resolve the entire set of molecules, severely limiting the amount of sequence information that can be obtained. The amount of sequence information that can be read, and the ease with which it is obtained, depend on both the gel system used and the way in which the electrophoresis is performed.

6.2 Overview

Sequencing gels are long, thin polyacrylamide gels which contain urea to denature the DNA and so reduce artifacts due to DNA secondary structure, and a buffer to allow electrophoresis. Polyacrylamide gels are formed by the polymerization of acrylamide and N, N'-methylenebisacrylamide to form a molecular sieve. The labeled DNA strands synthesized in the sequencing reactions are forced through this sieve by electrophoresis. This sorts them by size; the smallest molecules are retarded less by the gel and so migrate the fastest. After a suitable period of this electrophoretic separation, typically 3–10 h, the gel is usually transferred from the glass plates which supported it during electrophoresis to a sheet of filter paper and then dried. The dry gel is then exposed to X-ray film for autoradiography, typically for 12–24 h.

6.3 Reading a sequence autoradiogram

Following separation by denaturing gel electrophoresis, the labeled DNA strands are detected, for example by autoradiography. Fragments terminated near the primer will be shorter and move faster through the gel. The autoradiogram is therefore read from bottom to top, which gives sequence from 5′ (nearer the primer) to 3′ (further from the primer). This is illustrated in *Figure 6.1*.

If more than one band is seen corresponding to a single base, then the identity of that base cannot be determined unambiguously. This may be due to an artifact of sequencing (see Chapter 8), or may represent a mixed template with more than one base at a given position (see Chapter 10). In either case, the data needs to be recorded. The International Union of Biochemistry (IUB) and the International Union of Pure and Applied Chemistry (IUPAC) have defined a standard set of single letter codes to represent mixed bases [1]. This code is shown in Appendix B.

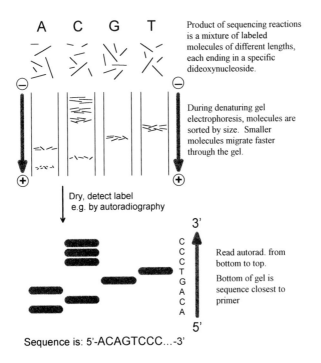

Sequence is: 5′-ACAGTCCC...-3′

FIGURE 6.1: Sequencing gels.

6.4 Gel systems

Sequencing apparatus for manual sequencing is available from a variety of manufacturers. I use Life Technologies' S2 system, which is relatively inexpensive, simple and reliable. The general parameters to consider are discussed below.

6.4.1 Safety

Sequencing gel systems use high-voltage power supplies (1500–3000 V). Safety legislation varies from place to place, but it is only sensible to ensure that the live parts of the apparatus are shielded from accidental contact. All commercially available apparatus that I have seen has some such safety feature built-in, but 'home-made' apparatus may not.

Arcing, leading to a fire risk, is another potential hazard. Leakage from the top buffer tank may form a short-circuit, or may lead to local heating. These are not just theoretical hazards, and it is essential to check for leakage from the top buffer tank every time a gel is run, and to consider these hazards before leaving a sequencing gel to run unattended for long periods, especially overnight.

6.4.2 Gel plates

Gel plates are normally made of glass. They need to be washed and dried carefully between each run, and this handling means that they will inevitably chip and break. Careful handling and storage will minimize this, but they will be cheaper to replace in a system that uses a simpler design, rather than one which has a more complicated shape. For example, the S2 system uses plates that are simple rectangles, and so are particularly cheap to replace. Indeed, a glass-cutter should be able to produce satisfactory replacements for a very modest cost.

6.4.3 Combs

Shark's-tooth combs allow more wells in the gel than conventional combs and allow more sequence to be read as the bands are adjacent to each other. With practice, they are better in every respect. The combs must be exactly the same thickness as the spacers. These combs are quite fragile, and prone to distortion if mistreated, but with care will last indefinitely. Mistreatment includes uneven pressure when inserting or removing the comb from the gel and not cleaning and storing the combs carefully. Combs for the S2 apparatus are provided with two different well widths, allowing 48 or 96 wells per gel. I routinely use the double-fine combs (96 wells/gel) to maximize throughput, but the wider combs give wider bands on the autoradiogram that are slightly easier to read.

6.4.4 Width

Narrow gels are easier to handle, but a wide gel (30–40 cm) allows more samples per gel, increasing throughput. Buffer gradient gels [2] are much harder to pour when using these wide gel systems.

6.4.5 Thickness

Gels are typically 0.4 mm thick. This is a compromise between the better resolution of thinner gels and the greater difficulty in handling them. The thickness depends on the spacers. It is of course essential that the spacers and combs are all precisely the same thickness.

Wedge gels are not recommended. They give more even spatial distribution of bands than conventional gels by reducing the voltage gradient at the bottom of the gel, but they take much longer to dry than conventional gels and do not seem to resolve the bands so well. Gels for automated sequencers do not have to be removed from the gel apparatus for autoradiography, and so can be made thinner. Gels for the ABI 377 are 0.2 mm thick.

6.4.6 Length

Sequencing gels are typically 40–50 cm from top to bottom. This is a compromise between the better resolution of longer gels and the difficulty in handling them. Gels up to 100 cm long can be run on some sets of apparatus (e.g. S1 system, Life Technologies), but pouring, fixing, drying and exposing these gels becomes more difficult.

6.4.7 Temperature control

Electrophoresis tends to set up a temperature gradient across the gel. This leads to uneven band mobility across the gel, giving an effect known as 'smile'. This gradient can be dissipated by attaching a metal plate to the gel plates (as in the S2 system), or using the buffer reservoir as a heat sink. An active cooling system such as a fan will allow higher power to be used, giving shorter run times. A thermostat system allowing accurate control of the gel temperature will aid reproducibility, and is an advantage when using Long Ranger™ gels (Section 6.5.1), which need to be run at 40–50°C.

6.5 Reagents

To obtain the highest quality sequence information, it is essential to use reagents of the highest possible quality for the gel. For safety, reproducibility and convenience, I recommend the use of liquid acrylamide mixes designed for sequencing (e.g. Sequagel, National Diagnostics). If solid acrylamide is used, fresh solutions should be made at least weekly and stored in the dark. One advantage of the liquid pre-mixes is that there is no risk of contamination by acrylamide powder. Acrylamide is a potent neurotoxin in either solid or liquid form.

6.5.1 Long Ranger™

A modified acrylamide solution with improved properties is available (Long Ranger™, FMC). This allows somewhat more sequence information to be determined in a single run, particularly if used with a discontinuous buffer system. Furthermore, the gels run substantially faster, minimizing the run time. Long Ranger gels are also said to be tougher and, therefore, easier to handle. In my experience, however, while the gels are more tear-resistant, they are much more prone to stretch and distort, even after cooling, so overall I find them harder to handle. Furthermore, the temperature of the gel needs to be maintained fairly accurately at 40–50°C during the run. In common with most sequencing apparatus, my S2 apparatus does not have a thermostatic control, so I have attached temperature-sensitive strips (Pharmacia, FMC) to the plates to monitor the temperature. I feel that the advantages of the Long Ranger reagent considerably outweigh the disadvantages, and so I now use it routinely. Long Ranger is also suitable for use in automated sequencers. Long Ranger gel solution is essentially used as a direct replacement for acrylamide solution; *Protocols 6.1* and *6.2* note the minor differences between the two.

Protocol 6.1 uses 1 × Tris–borate–EDTA (TBE) in the gel and running buffers. This can be modified for other applications. For maximum

PROTOCOL 6.1: Preparing and pouring a sequencing gel

Materials required
- sequencing gel system
- silanizing agent (BDH, FMC)
- plastic film (e.g. Saran wrap)
- *either* Long Ranger ™ gel solution
  or commercial acrylamide concentrate
  or 'home-made' 40% 19:1 acrylamide concentrate: per liter: 380 g acrylamide, 20 g *N, N´*-methylenebisacrylamide, store in dark and use within 1 week. *Caution: acrylamide is a potent neurotoxin*
- urea
- 10% ammonium persulfate (store at 4°C for up to 1 week)
- TEMED (*N, N, N´, N´*, -tetramethylethylenediamine)
- 10 × TBE: per liter: 108 g Tris base, 55 g boric acid, 40 ml 500 mM EDTA pH 8.0
- 50 ml syringe (optional)

Method
Different gel systems may require different volumes of gel mix, or have their own gel casting apparatus or methods. This protocol is based on Life Technologies' model S2, a typical system with two glass plates, 0.4 mm spacers and shark's-tooth combs.

A. Preparing the gel plates
1. Wash plates thoroughly with warm soapy water. Make sure that all traces of dried gel and grease are removed. Rinse well with distilled water. Wash with ethanol and wipe dry.
2. Silanize one plate to ensure that gel can be removed easily after electrophoresis. Ensure that the same plate is silanized every time or the gel will soon stick to neither. I always silanize the short plate.
3. Assemble plates into a mold according to the manufacturer's instructions. For the S2, I seal around the sides and bottom with 4 cm adhesive tape, carefully folding the tape around the edges and corners to ensure a good seal. Some systems need a very small amount of petroleum grease to seal the corners, but with this method there should be no need. Clamp the edges of the plates with bulldog clips positioned directly over the spacers. Support the plates at an angle of about 30° from the horizontal and with a slight tilt from one side to the other. Make a reservoir at the lower side of the open end by taping across from one spacer 'ear'.
4. Cover the entire working area with absorbent paper. I rarely manage to pour a gel without dripping some gel solution onto the bench.

B. Pouring the gel
1. Combine in a clean conical flask or beaker:
   - Long Ranger gel solution   7.5 ml
   - 10 × TBE   7.5 ml
   - urea   31.5 g
   - Deionized water   to 75 ml

   Mix by gentle swirling. The mix can be warmed to dissolve the urea, but should be cooled to room temperature before proceeding. This mix gives 75 ml of 5% Long Ranger in 1 × TBE. S2 gel volume is 60 ml, so this allows some leeway for leakage/spillage. If using 40% acrylamide mix instead of Long Ranger, use 9.4 ml to give a final concentration of 5% and reduce the buffer concentration to 0.6 × TBE (see text). De-gas 'home-made' solutions under vacuum.
2. (Optional) Filter solution through filter paper or Nalgene cellulose acetate filter (0.45 μm).
3. Add:
   - TEMED   37.5 μl
   - 10% ammonium persulfate   375 μl
4. Slowly and carefully pour the gel solution into plates. I pour direct from the conical flask, others prefer to use a 50 ml syringe. When the gel is nearly full, tilt the gel almost to horizontal. This reduces its capacity so that the mold should be full with a little excess at the top. Try not to allow bubbles to enter the gel. This will be minimized by pouring slowly and continuously. Dirty gel plates are the other major cause of bubbles. Bubbles can sometimes be dislodged with a fine wire.
5. Insert the comb, flat side about 3 mm into the gel if using shark's-tooth combs, otherwise with the serrated side fully into the gel. Clamp the comb in place to ensure an accurate fit after polymerization.
6. The gel should set in about 30 min, but should be left longer, preferably overnight, to allow complete polymerization. After initial polymerization, lay damp paper towels over the open end and wrap with plastic film. This is to prevent the top of the gel from drying out.

PROTOCOL 6.2: Running a sequencing gel

Materials required
- sequencing gel system
- high voltage (2000–3000 V, 50–60 W) power pack
- gel dryer
- vacuum pump (or water pump) with vapor trap
- Temperature monitoring strips (FMC)
- 1 × TBE (see *Protocol 6.1*)
- (optional) gel fix: *either* 10% methanol, 10% acetic acid *or* 20% ethanol, 10% acetic acid

Method
Different gel systems may require slightly different methods. This protocol is based on Life Technologies' model S2, a typical system with two glass plates, 0.4 mm spacers and shark's-tooth combs.

A. Pre-running and loading the gel
1. Starting with a prepared gel (from *Protocol 6.1*), remove the plastic film, paper towels bulldog clips and tape, being careful not to dislodge the spacers. Clean the outside of the gel plates if necessary. Mount the gel plates in the electrophoresis apparatus according to the manufacturer's instructions and add the appropriate amount of running buffer (e.g. 1 × TBE, see text). Remove the combs, first noting their location and orientation (see step 5 below).
2. Pre-run the gel for 10–15 min. Use 48–55 W for the S2, adjust appropriately for larger or smaller gels (e.g. 28–35 W for 40 cm × 20 cm × 0.4 mm gels). If using acrylamide rather than Long Ranger, use about 1.5 × this power.
3. Denature the sequencing reactions by heating to 80–90°C for 2–3 min and chill rapidly on ice to prevent reannealing.
4. After the pre-run, use a syringe to wash the top of the gel thoroughly with running buffer, to remove urea.
5. Replace the shark's-tooth combs, serrated side down. The teeth should touch or just penetrate the top of the gel. Combs may not be of exactly uniform thickness so, to ensure a perfect fit, replace combs in their original position and orientation (except that the teeth are down instead of up).
6. Load the gel by pipeting 0.5–1.0 μl of denatured sequencing reaction into each well. Load sets of four reactions into adjacent wells in a standard order, e.g. ACGT. Special flattened pipet tips are available to facilitate this; I flatten the end of a standard tip. Load the samples as rapidly as possible to minimize diffusion of the sample and of urea into the wells.
7. Run the gel. Adjust the power so that the gel temperature is 40–50°C. The gel temperature can be monitored by sticking temperature-sensitive strips to the gel plate. If using acrylamide rather than Long Ranger, the gel temperature should be higher (~60°C) and is less critical. The power recommendations in step 2 above are intended to give these temperatures, but should be adjusted for your apparatus.
8. Multiple loading can give much more sequence information from a single set of reactions. I run the bromophenol blue (dark blue) marker dye to the bottom of the gel, continue running for 30 min and then stop the gel. Remove the comb, wash out the next set of wells, as in step 4, and repeat steps 5–7. Run time is approximately 4 h to read 400–500 bases from the primer. The equivalent run time for acrylamide gels is 6–8 h. As a guide, in a 5% Long Ranger gel, bromophenol blue co-migrates with 40 base fragments, xylene cyanol with 175 base fragments (35 and 130 respectively for acrylamide).
9. When the run is complete, turn off the power supply and remove the plates from the electrophoresis apparatus.

B. Autoradiography
1. Allow the gel to cool to near room temperature. Lay the gel down with the silanized plate on top. Insert a scalpel or spatula between the plates and carefully separate them. The gel should stick to the lower (un-silanized) plate only. Remove the spacer and combs.
2. (Optional) Fix the gel and remove urea by soaking the gel in gel fix for 20 min. The gel should remain attached to the glass plate during this process. This step is essential for formamide gels.
3. Transfer the gel to a sheet of Whatmann 3MM filter paper. Cover the gel with plastic film (e.g. Saran Wrap) and dry under vacuum at 70–80°C for 30–60 min. Fixed gels dry slightly faster.
4. Remove the plastic film and expose to X-ray film. Make sure that the gel is pressed flat against the film. Typical exposures are 24–48 h.

read lengths, use $0.6 \times$ TBE in the running buffer and $1.2 \times$ TBE in the gel. However, the top of the gel degrades after a few hours under these conditions, which effectively prevents multiple loading. For a large number of samples, I recommend making two gels and loading each sample onto both gels. One is run for a short period and one for an extended period. This gives the same information as loading the sample twice on the same gel, but allows the use of the optimum buffer conditions. For fast runs, to read only 200–300 bases, use a 5 or 6% gel with $0.6 \times$ TBE in both the gel and the running buffers.

6.5.2 Glycerol-tolerant gels

High concentrations of glycerol in the sequencing reaction improve the temperature stability of the T7 DNA polymerase. This allows the reaction to be performed at a higher temperature, potentially reducing effects of secondary structure (see *Figure 3.5*). A final glycerol concentration of up to 50% can be used, compared with the 0.8% in the standard reaction. However, in the standard TBE buffer system, glycerol interferes with the gel electrophoresis, distorting bands at more than about 350 bases from the primer. This effect is caused by a reaction between borate and glycerol, and can be eliminated by replacing the borate with taurine (Amersham/USB). Use TTE (89 mM Tris, 28.5 mM taurine, 0.5 mM EDTA, pH 8.0) in place of TBE. Alternatively, this problem can be eliminated simply by washing out the wells once the sample has entered the gel.

6.5.3 Formamide gels

'Compressions' are a major source of error in analyzing sequencing gels. A compression is caused by intramolecular base pairing leading to a folded or hairpin structure (see Section 8.2). Such a structure may migrate faster through the gel than an equivalent linear molecule, so the even spacing of the bands is disrupted. One method of eliminating gel compressions is to make the gel conditions more denaturing by adding formamide in addition to urea (see *Protocol 6.3*). Formamide is typically used at a final concentration of 40–50%.

6.5.4 Capillary electrophoresis

Acrylamide is not necessarily the best gel matrix for high-resolution denaturing gel electrophoresis [3–5]. Improvement in the separation could lead to better separation and longer reads (more sequence information per sample). Liquid polymers in a capillary offer a possible improvement. Such a system would probably require real-

---

PROTOCOL 6.3: Formamide gels

Materials required
As detailed in *Protocols 6.1* and *6.2,* with the addition of
• formamide

Method
As described in *Protocols 6.1* and *6.2* with the following modifications.

A. *Pre-running and loading the gel (see* Protocol 6.1*)*
1. Increase the concentration of Long Ranger or acrylamide to 8%.
   Add formamide to 40% (30 ml) while keeping the final volume at 75 ml.
   This combination results in a gel of similar performance to a standard 5% gel.
2. Increase the amounts of TEMED and ammonium persulfate to:
   • TEMED   60 µl
   • ammonium persulfate   400 µl

B. *Running a sequence gel (see* Protocol 6.2*)*
1. Formamide gels run about half as fast as standard gels. The running time should therefore be doubled.
2. Formamide gels must be fixed prior to drying.

---

time detection as the matrix cannot be extracted intact, unlike an acrylamide gel. The ABI 310 uses a single capillary with an internal diameter of 75 µm. This has the advantage of a short run time but the disadvantage that it can only handle a single sample at a time. Automated sequencers are described in Chapter 7.

References

1. IUPAC–IUB Commissions on Biochemical Nomenclature (CBN) (1970) Abbreviations and symbols for nucleic acids, polynucleotides and their constituents, recommendations. *Eur. J. Biochem.*, 15, 203–208.
2. Sheen, J.-Y. and Seed, B.B. (1988) Electrolyte gradient gels for DNA sequencing. *BioTechniques*, **6**, 942–944.
3. Drossman, H., Luckey, J.A., Kostichka, A.J., D'Cunha, J. and Smith, L.M. (1990) High-speed separations of DNA sequencing reactions by capillary electrophoresis. *Anal. Chem.*, **62**, 900–903.
4. Luckey, J.A., Drossman, H., Kostichka, A.J., Mead, D.A., D'Cunha, J., Norris, T.B. and Smith, L.M. (1990) High speed DNA sequencing by capillary electrophoresis. *Nucleic Acids Res.*, **18**, 4417–4421.
5. Swerdlow, H., Zhang, J.-Z., Chen, D.-Y., Harke, H.R., Grey, R., Wu, S., Dovichi, N.J. and Fuller, C. (1991) Three DNA sequencing methods using capillary gel electrophoresis and laser-induced fluorescence. *Anal. Chem.*, **63**, 2835–2841.

7 Nonradioactive Methods

7.1 Introduction

The sequencing methods described in Chapters 3 and 6 are based on the incorporation of a radiolabel into the synthesized strand. Although this is still by far the most widely used approach, there are several nonradioactive alternatives available. Semi-automated sequencing machines, which detect and analyze fluorescently labeled DNA with sophisticated detection systems and analysis software, are discussed separately below. Most other methods closely resemble the standard method, but incorporate a nonradioactive label into the DNA, for example biotin (Amersham) or digoxigenin (Boehringer Mannheim). Unlike ^{35}S, these haptens cannot be detected directly, and so the DNA must be transferred from the sequencing gel to a solid support, for example a nylon membrane, and then detected using an enzymatic detection system [1–3]. Modifications allow the use of multiple labels in the same lane, which are then detected sequentially to allow several sets of sequence information to be read from a single set of lanes on a gel [4, 5]. An alternative is Promega's Silver Sequence™ system, in which the sequencing gel is stained to reveal the bands directly. A hard copy of the result can be made using a special film (Kodak EDF), which will be essential for most *de novo* sequencing projects, where old results need to be re-examined to resolve discrepancies.

The advantages of these systems are:

- they do not require the use of radioisotopes;
- the time from running the gel to getting the sequence information is generally reduced, as the 12–24 h autoradiography step is eliminated.

The disadvantages are:

- the data obtained are rarely of such high quality as those obtained by autoradiography;
- the 'hands-on' time is generally increased by the additional manipulations required.

As a consequence, nonradioactive methods other than semi-automated sequencers are rarely used. One system that requires further consideration is the GATC® 1500 (GATC GmbH). These systems automatically transfer the electrophoretically separated reaction products on to a nylon membrane by moving the membrane past the bottom of the gel. In this way, every molecule has passed through the entire length of the gel, optimizing the separation. Furthermore, the speed at which the membrane moves past the bottom of the gel can be regulated to control the band separation. These innovations potentially allow much more sequence information to be obtained from a single run – well over 1000 bases under ideal conditions. These sequencing systems are more expensive than a conventional gel apparatus, but still reasonably affordable for many laboratories, and much cheaper than a semi-automated sequencer.

7.2 Semi-automated sequencers

The development of instruments which automate data collection from a sequencing gel run has dramatically increased the throughput of sequencing laboratories. These instruments are called automated or semi-automated sequencers. Unlike the standard manual methods, which use a radioactive label and visualize the bands by autoradiography, automated sequencers use sensitive photodetectors to detect DNA fragments labeled with fluorescent dyes as they pass a scanning laser near the bottom of the gel [6, 7]. This automated data collection is linked to analysis software to automate the entire process from the point of loading the sequencing reaction on to the gel.

Most new sequence is determined using semi-automated sequencers. This is not because most laboratories have one, but simply that the genome sequencing projects rely on the high throughput and reduced operator time of these instruments. This high throughput is the basis of the genome sequencing projects, which have recently determined the entire sequence of *Saccharomyces cerevisiae* (completed May, 1996) and will, within the next few years determine the complete sequence of a number of other organisms of scientific, agricultural or medical significance, including that of *Homo sapiens*. The genome sequencing projects have dramatic implications for future sequencing projects, as discussed in Chapter 11.

Semi-automated sequencing machines are available from several manufacturers. They all share one major disadvantage – their price of $60 000 – $150 000 puts them out of consideration for most individual laboratories. However, some departments are now purchasing them to establish a central facility, so an increasing number of researchers now have access to such a machine. This is in many ways the ideal solution, as the sequencer really needs to be operated by a single, trained individual, to ensure the extreme care and cleanliness that these sensitive instruments require.

Detailed operating instructions for any of these instruments are beyond the scope of this book, and the necessary training material is available from the manufacturers. I will briefly describe the principles behind two popular sequencers – the ABI 377 and the LI-COR – and summarize their advantages and disadvantages relative to conventional sequencing systems.

7.3 ABI 377

The ABI 377 can simultaneous detect fluorescence at four different wavelengths, set to coincide with the emission of four different fluorescent dyes. This simultaneous detection is performed by separating the emitted light with a diffraction grating, allowing simultaneous detection of multiple wavelengths with a charge-coupled device (CCD) camera. Each dye is used to label strands terminating in one of the four possible dideoxynucleotides. The reaction mix is run in a single lane so that the color of each band passing the detector represents the DNA sequence (*Figure 7.1*).

A key advantage of automated sequencing is the automated data collection. The ABI 377 illuminates the gel with an argon laser and monitors the fluorescent emissions from the labeled DNA. This real-time data collection means that there is no opportunity to adjust for variable signal strength, unlike manual sequencing where an autoradiograph can be exposed for longer if necessary. Sequencing reactions must, therefore, be carefully optimized (see below). Furthermore, the faint signals require extremely sensitive detectors, which leads to the system being very sensitive to even minute levels of contamination by extraneous fluorochromes. Such contaminants may come from seemingly innocuous sources such as paper towels, to which fluorochromes are added to enhance their color.

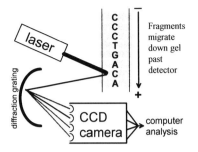

FIGURE 7.1: *Real-time data collection on an ABI 377. As each fragment moves past the detector, the base-specific dye is excited by the laser. The dye emits light at its characteristic wavelength. This fluorescent emission is separated according to its wavelength by a diffraction grating. The signal is detected by a sensitive charge-coupled device (CCD) camera, and recorded and analyzed by computer. The software separates the signal into four channels, each corresponding to one of the dyes. Use of a diffraction grating allows all four channels to be monitored simultaneously.*

At first sight, the raw data from an ABI 377 do not look very encouraging. Each lane is a ladder of colored bands, each representing a single nucleotide. However, the resolution and relative intensities of these bands appear to be at least no better than conventional sequencing gels. This raw data is processed by the associated software to give the best-known output, called an electropherogram (see *Figures 7.2* and *7.3*). This is analyzed further by the assignment of values to each peak, so that the final output is the DNA sequence in text format. The base-calling software is very good, but manual editing can generally improve the accuracy slightly, particularly towards the end of the readable sequence, where single guanine residues may be so faint as to be missed, especially near adenines, and homopolymeric tracts may be miscounted. The ability of the analysis software to interpret the raw data ultimately accounts for the superior performance of the ABI automated sequencers compared with manual sequencing, but there are also some inherent advantages and disadvantages.

The sequencing reaction is loaded onto a single lane of the gel. The four termination reactions are each labeled with a different color so, unlike other methods, there is no need to run a set of four lanes for each reaction. This eliminates the need for mobility comparisons between different lanes, and also allows a large number of samples to be run on the same gel, maximizing throughput. Up to 64 sets of sequencing reactions can be run simultaneously. However, the use of

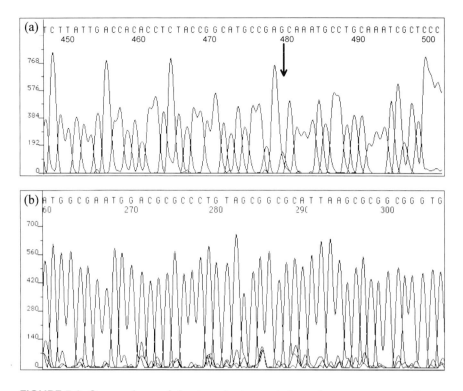

FIGURE 7.2: *Comparison of dye terminator and dye primer sequences. Dye terminator chemistry (a) gives much greater variation in peak height than dye primer chemistry (b). Low G peaks are a common problem, particularly following an A residue. An example of such a low G peak is arrowed in (a).*

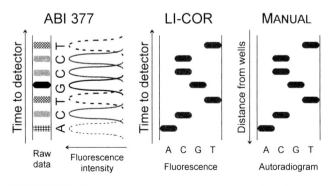

FIGURE 7.3: *Correspondence between manual and automated sequence data collection.*

four different dyes has an inherent problem. Addition of a dye will affect the mobility of a DNA fragment. Since the four nucleotides are labeled with different dyes, their mobilities are not directly comparable. In practice, this appears to be only a minor problem, which can be compensated for by the gel analysis software, but it does limit the range of labeled primers that can be used to those with known electrophoretic effects.

Another major advantage of automated sequencers over conventional gels is that the sequence is read at a fixed point of the gel, rather than after a fixed period of time. This means that the sample has been resolved optimally by electrophoresis through the entire length of the gel (36 or 48 cm well-to-read length for the ABI 377). In conventional gels, the electrophoresis is stopped after a pre-determined time, the gel is processed and the sequence ladder detected, typically by autoradiography. Sequence at the bottom of the gel is well resolved, but most of the sequence is higher up the gel and is less well separated.

There are two chemistries available for the ABI 377: dye primers and dye terminators. The difference is simply the point at which the fluorescent label is incorporated into the newly synthesized strand. Each method has its advantages and disadvantages, so that it is important to choose the right method for a given application.

7.3.1 Dye terminator chemistry

In dye terminator chemistry, each of the four dideoxynucleotides is labeled with a different fluorochrome [8, 9]. Incorporation of a ddNTP therefore simultaneously terminates chain elongation and labels the terminated molecule. A corollary of this is that chains terminated for other reasons, for example template secondary structure, impurities, etc., are not labeled and so are not detected. This makes this chemistry less prone to 'false stops'. This can be exploited further – incorporation of dITP in the reaction mix reduces compressions, but increases false stops. Since false stops are not detected, this allows the routine use of dITP in the reaction mix, greatly reducing the number of sequencing errors due to compressions. Another advantage of dye terminator chemistry is that only a single reaction need be performed. Since a different label is incorporated with each ddNTP, there is no need to perform four separate reactions, one for each ddNTP, as there is in conventional sequencing reactions.

Dye terminators can be used with either cycle sequencing or 'one-pass' sequencing protocols. Of these, cycle sequencing is by far the more common. The principal advantages are the low molar amounts of template DNA required, which is important when sequencing PCR products or high molecular weight templates, and the convenience. Using a commercial kit, with all the reagents pre-mixed (ABI PRISM Ready Reaction kits, Perkin-Elmer), performing the sequencing reaction simply requires the mixing of DNA, primer and water with an aliquot of the pre-mixed reagents from the kit, followed by 20 cycles in a thermal cycler. The unincorporated labeled nucleotides are removed by ethanol precipitation and the sample air-dried. At this point, the reaction is stable for up to a week at −20°C. Just prior to gel electrophoresis, the sample is resuspended in a gel loading buffer and heated to denature the DNA.

Various claims are made for typical and maximum read lengths using this system. For a typical double-stranded plasmid template, I expect to be able to read about 450–500 nucleotides at 99.8% accuracy, that is no more than one error (ABI 377, 36 cm well-to-read gels). This is certainly more than I can obtain by manual sequencing. ABI suggest that 650 bases is typical, but this is at 98% accuracy, which I consider to be too low. An accuracy of 98% over 650 bases implies 13 errors. These will be concentrated in the last part of the sequence, so I prefer not to read this at all. An alternative strategy, appropriate in some cases, is to accept this lower quality data on the basis that it will be corrected by re-sequencing the region from other primers and on the other strand. It is certainly true that *differences* between two sequences can be determined much further from the primer than 500 bases, if the two templates are sequenced in parallel and then the electropherograms compared.

Wild-type *Taq* DNA polymerase, the most commonly used enzyme for PCR, discriminates very well between dNTPs and ddNTPs, so that it incorporates ddNTPs, and particularly labeled ddNTPs, only very inefficiently. This means that high concentrations of labeled ddNTPs have to be used. This results in a substantial carry-over of unincorporated label onto the gel, unless it is carefully removed, for example by gel filtration. This is clearly undesirable, so Roche Molecular Systems, together with Perkin-Elmer, have developed a mutant *Taq* polymerase, AmpliTaq® DNA polymerase, FS. This enzyme has one mutation which virtually eliminates the 5′→3′ nuclease activity and a second (F667Y) which allows much more efficient incorporation of ddNTPs [10, 11]

With either version of *Taq* DNA polymerase, the frequency of incorporation of a labeled ddNTP at a given position is strongly dependent on the sequence context. This leads to the major disadvantage of dye terminator chemistry, which is the uneven peak

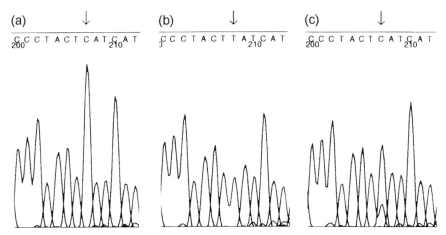

FIGURE 7.4: *Disparity in peak heights at the polymorphic site in a heterozygote. (a) Sequencing trace of the homozygote with the allele CTC. The 3′ C peak is large and is about twice the size of the C peak in a heterozygote. (b) Sequencing trace of the homozygote with the allele CTT. The 3′ T peak is small but is still twice the size of the T peak in a heterozygote. (c) Sequencing trace of the heterozygote with CT (C/T). The base-calling program called this heterozygous base a 'C' rather than an 'N'. Reproduced from [13] with permission from Eaton Publishing.*

heights on the resultant electropherogram. This reduces the maximum read length somewhat but, more seriously, it makes it almost impossible to recognize the presence of a mixture of bases at a given position. This makes dye terminator chemistry unsuitable for direct sequencing for mutation detection (see *Figure 7.4* and Section 10.3). The variable peak height is strongly sequence-dependent, the best known problem being that G peaks are very low when preceded by A peaks. The effect of sequence context on peak height depends on the enzyme (and terminators) used, and has been analyzed for both wild-type and mutant *Taq* [12, 13]. This analysis should lead to the development of better base-calling software.

7.3.2 Dye primer chemistry

In dye primer chemistry, the fluorescent dye is attached to the primer [14]. This means that, as in conventional sequencing reactions, four separate reactions are required, one with each labeled primer. Furthermore, four different primers are required for each set of four reactions. These each have the same sequence, but are conjugated to a different fluorochrome. Synthesis of a single fluorescently labeled primer is more expensive than of a single unlabeled primer. A set of primers for dye primer sequencing therefore costs more than four times as much as the equivalent primer for dye terminator sequencing. Furthermore, the dye interacts with the first five bases at the 5' end of the primer sequence. This affects the mobility of the dye-labeled sequencing fragments, though the effect is noticeable only for short fragments (< 50–75 nucleotides). This mobility shift is reproducible and can be compensated for by the gel analysis software *as long as* the first five bases of the primer match those of a standard primer. This can be achieved simply by adding the first five bases of the M13 reverse primer (5'-CAGGA-3') at the 5' end of the custom dye primer sequence, but of course this further increases the cost of the custom dye primer set. In practice, this means that dye primer chemistry is used only with standard primer sets – the 'universal' primers or else custom sets for specific applications but not, for example, for primer walking.

Dye primer chemistry also lacks two advantages that apply to dye terminator chemistry. The sequencing reactions must be performed in four separate tubes, then pooled for gel electrophoresis, unlike the single-tube reaction allowed by dye terminator chemistry, so a little more operator effort is required. Furthermore, all extensions from the primer are labeled, so premature terminations ('false stops') are detected, unlike dye terminator chemistry (see above). This in turn means that dITP cannot be used routinely, because it leads to an unacceptable level of false stops. 7-Deaza-guanine is used instead, but it is less effective, and so dye primer reactions are also more sensitive to compressions.

Dye primer chemistry has the single major advantage that the signal intensities are much more uniform from one base to the next. This is essential when the template may contain a mixture of bases at a given position, for example a heterozygous mutant or a virus population. A heterozygote will have two different bases at the same position, and this must appear as two superimposed signals of equal intensity, and half the intensity of a homozygous base. Dye primer chemistry with AmpliTaq® DNA polymerase, FS, or with Sequenase, achieves this, whereas dye terminator chemistry does not.

7.4 Optimizing sequencing on the ABI 377

In general, sequencing protocols for fluorescent automated sequencing are less robust than manual methods. This is partly due to the real-time data collection, which cannot be adjusted for poor signal strength. An additional problem is the sensitivity to contaminants of the polymerases used, particularly with dye terminator chemistry. The most common factors which diminish sequence quality are impure template or primer preparations, inappropriate template or primer concentrations and inefficient removal of unincorporated labeled dideoxynucleotides or primers.

7.4.1 Template quality

Fluorescent automated sequencing is far more sensitive to contaminants than is manual sequencing. This means that a number of template preparation methods that are perfectly satisfactory for manual sequencing give only erratic results with automated sequencing. I therefore invariably purify my templates using Qiagen columns: tip-20 for templates which will only be sequenced once or a few times, tip-100 for templates which will be sequenced repeatedly. Templates for single reactions include elements of a deletion series, whereas in a primer walking project the same template would be sequenced repeatedly. Other proprietary methods may also give satisfactory sequence, but in general silica-based methods (including Qiagen's version) seem less reliable than Qiagen's ion-exchange columns. Sequencing reactions are also sensitive to contamination with salt and ethanol, so it is essential that any ethanol precipitations are followed by a careful wash with 70% ethanol to remove precipitated salt, and are then carefully dried.

The quality of the initial culture from which the DNA is extracted can also affect plasmid template quality. Some host strains (HB101, TG1, TG2, JM100 series) release large amounts of carbohydrate upon lysis. This can interfere with the DNA purification. Furthermore, these *EndoA+* strains produce relatively large amounts of nuclease. While plasmid purification procedures should be capable of efficiently removing carbohydrate and protein from the DNA, it would be sensible to use host strains such as DH1, DH5α, C600 and XL1-Blue in preference. Promega's Wizard plus SV miniprep kits include an alkaline protease to inactivate endonuclease A.

The culture conditions can also have an effect. Cultures left at stationary phase for extended periods will begin to lyse, releasing degraded genomic and plasmid DNA. It is also important that antibiotic selection is applied for as long as possible during the culture growth. This is a particular problem with ampicillin, which is inactivated by plasmid-encoded β-lactamase. This secreted enzyme is present at high concentration in stationary phase cultures. New cultures should, therefore, be innoculated with a fresh colony from a plate, rather than stored liquid cultures. Both these considerations –

lysis in stationary phase and ampicillin degradation – suggest that overnight cultures should be grown in LB media rather than rich media such as TB [15]. The slower bacterial growth in LB means that the culture will reach stationary phase in 10–12 h, rather than around 8 h for TB. A typical overnight culture is grown for 16 h, so a TB culture will normally have more lysed cells than a comparable LB culture (*Figure 7.5*).

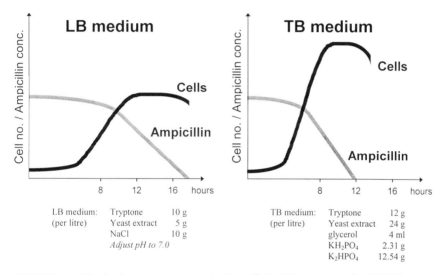

LB medium: (per litre)	Tryptone	10 g
	Yeast extract	5 g
	NaCl	10 g
	Adjust pH to 7.0	

TB medium: (per litre)	Tryptone	12 g
	Yeast extract	24 g
	glycerol	4 ml
	KH_2PO_4	2.31 g
	K_2HPO_4	12.54 g

FIGURE 7.5: *Typical growth curves for* E. coli *in different media [13]. For a typical overnight culture of 16 h, a TB culture will contain more lysed cells. Furthermore, the culture will have been without selection for longer, due to inactivation of the ampicillin by plasmid-encoded β-lactamase.*

7.4.2 Primer quality

It is obviously essential that the large majority of the oligonucleotides in the primer are of the desired sequence. With modern synthesis methods, this should not be a problem for primers in the typical size range of 18–25 nucleotides. Opinions vary, however, as to the importance of post-synthesis purification. I have used crude desalted oligonucleotides from a number of commercial suppliers. These are usually perfectly satisfactory, but occasionally fail. This failure is primer-specific as the same template can be sequenced satisfactorily with other primers. Surprisingly, I find that these poor primers may work on some templates but not others. The reasons for this are not clear, but it seems to be a general problem. Other groups avoid this difficulty by using more highly purified oligonucleotides, for example HPLC-purified. I currently consider that the rate of failure is low enough to tolerate, in view of the extra costs associated with further purification. As synthesis methods continue to improve, it seems certain that the price and quality of oligonucleotides will improve further.

7.4.3 Template and primer concentrations

Signal strength needs to be optimized for automated sequencers because of the real-time detection. This means that the template and primer concentrations in the sequencing reaction have much tighter limits than for manual sequencing. Template DNA concentrations should be determined spectrophotometrically or by careful comparison of ethidium bromide samples with known standards. Too little template will give weak signals. Too much template will rapidly deplete the nucleotides in the reaction mix, leading to an excess of short molecules and premature termination. Primer concentrations are normally determined by the manufacturer but can be measured spectrophotometrically if necessary. Qiagen has determined optimum template amounts for a standard 20 μl reaction with 3.2 pmol of primer (*Table 7.1*).

7.4.4 Removing unincorporated label

In standard ^{35}S sequencing protocols, the unincorporated label is a thionucleotide. This has a much higher mobility in the sequencing gel than the incorporated label and so runs off the bottom of the gel. Fluorescent sequencing with dye terminator chemistry uses dideoxynucleotides conjugated to a fluorochrome. These have a gel mobility comparable with the labeled strands, so unincorporated label will, if not removed, result in large, spurious peaks of fluorescence superimposed on the sequence ('dye blobs'). Unincorporated label can also cause the gel analysis software to start reading too early, or to scale the true sequence peaks incorrectly.

Unincorporated label can be removed by gel filtration, organic extraction (phenol/chloroform) or ethanol precipitation. As described above, AmpliTaq® DNA polymerase, FS allows the use of relatively low concentrations of labeled dideoxynucleotides. With this enzyme, ethanol precipitation is normally sufficient to remove the unincorporated label. Unmodified *Taq* requires higher concentrations

TABLE 7.1: Recommended amounts of template DNA for ABI 377 with various chemistries [13]

	Taq DNA polymerase		T7 DNA polymerase	
	Dye primer	Dye terminator	Dye primer	Dye terminator
Plasmids (3–10 kb)	0.8 µg	0.8 µg	5 µg	5 µg
Plasmids (10–20 kb)	1.2 µg	1 µg		
Cosmids (30–45 kb)		2 µg	n/a	n/a
P1 (80–100 kb)		3 µg	n/a	n/a
PCR products		0.1 µg/kb		
Single-stranded DNA	0.3 µg	0.2 µg	2 µg	2 µg

of label and so more efficient methods for removing unincorporated label are required. Apparently, removing negative charge by treatment with phosphatase will prevent fluorescent ddNTPs migrating into the gel, hence solving the dye blob problem.

7.5 Future developments

The advent of sequencing instruments capable of real-time detection and automated base-calling has allowed a huge increase in the rate at which sequence can be determined. The continuing demand for more accurate, longer sequencing reads and greater automation and throughput ensures that manufacturers will continue to develop and improve sequencing technologies. The main pressures for improvement come from two directions. Firstly, the genome sequencing projects need to sequence billions of bases with extremely high accuracy and at a low cost. Secondly, diagnostic applications need to detect heterozygotes efficiently while minimizing the false positive rate.

The likely directions for future improvement appear to be those discussed below.

7.5.1 Brighter dyes

Brighter fluorescent dyes would allow the use of less template DNA, or give stronger signals in an equivalent reaction. This would be particularly helpful for sequencing large templates such as cosmids or P1 phage, where the amount of template is limiting. Amersham have taken a step in this direction with their DYEnamic™ primers, which have two fluorescent dyes attached to the same primer molecule [16]. The donor dye (FAM) absorbs light and transfers the energy to the acceptor dye (REG, FAM, TAMRA or ROX), which radiates it at its characteristic wavelength. Amersham claim that the effective signal strengths are typically six times greater than for primers carrying single dyes. These energy transfer primers should soon be available from other commercial sources.

7.5.2 Better electrophoretic resolution

Gel matrices other than acrylamide may allow better electrophoretic separation, allowing longer (and perhaps faster) reads. Current favorites are liquid polymers. These are generally used in a capillary format, but this is difficult to scale up to a multi-lane, high-throughput instrument. ABI are considering a system of grooved plates, giving an array of pseudo-capillaries in a more conventional format.

7.5.3 Better software

Base-calling software is already very good. However, manual inspection of an electropherogram usually shows a few errors which are obvious to the operator, so the software could clearly be improved. Analysis of sequence-dependent peak height variation should allow some improvement [12, 13]. This might ameliorate the peak height variation seen with dye terminators.

7.5.4 Uniform peak heights

Dye terminator chemistry gives large variations in peak heights. This causes errors in base-calling, especially in the detection of heterozygotes. While it may prove possible to normalize the peak heights with more sophisticated software, it would be more satisfactory to eliminate the problem where it occurs, in the sequencing reactions. This may be possible by further modifying the polymerase to reduce the sequence dependence of dye–ddNTP incorporation. Changing the dyes on the terminators may also have an effect. Solving this problem would be a significant advance. New, improved dye terminators should become available soon.

7.5.5 Increased throughput

The obvious way to increase throughput is to increase the number of lanes per gel. This requires improved software, to distinguish accurately one lane from the next. A recent upgrade for the ABI 377 (377XL) increases the maximum number of lanes on a single gel from 36 to 64.

All stages from template preparation to gel loading could benefit from greater automation. Programable robots and dedicated devices are available to perform some of these tasks, but at a cost which makes them uneconomical for most laboratories.

7.6 LI-COR

Like the ABI instruments, the LI-COR range of automated sequencers detect fluorescently labeled DNA molecules as they are excited by a laser near the bottom of the gel. The fundamental advantages and disadvantages of real-time detection therefore apply equally to these instruments. There are, however, significant differences in detail.

The LI-COR instruments use a scanning fluorescence microscope to detect the fluorescent label [7]. This allows the detection head to focus on a single plane within the gel. This focusing can be controlled automatically, based on the differential fluorescence of the gel relative to the glass plates. There is only a single detection wavelength, so data can be collected more rapidly during each scan. On the other hand, this means that four lanes are required per sample, just as with radioactive labels (*Figure 7.6*).

Labeled primers are available, but the standard labeling system is internal labeling using dATP labeled with an infra-red dye. This is exactly analogous to labeling with [α-^{35}S]dATPαS, and so allows well-known manual or cycle sequencing protocols to be used with only minor modifications.

The LI-COR 4000LS system uses 66 cm plates as standard (though shorter plates are available), compared with 36 or 48 cm for the ABI 377. This is a trade-off between throughput and resolution – longer gels maximize band resolution but take longer to run than shorter gels. Compared with the ABI 377, the LI-COR range sacrifices throughput for read length. The LI-COR 4000LS system can give more than 800 bases per sample at 99% accuracy, but can only run seven or 11 samples per gel compared with 36 or 64 on the ABI 377. Longer read lengths can save time and money in primer walking strategies (Section 11.3) as less steps and fewer custom primers are

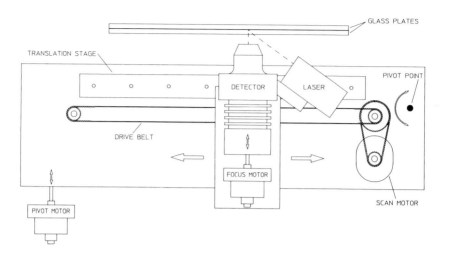

FIGURE 7.6: *Schematic presentation of the LI-COR scanning platform. The scan motor moves the fluorescence microscope/focus motor assembly back and forth on the translation stage via the drive belt. The focus motor moves the fluorescence microscope toward or away from the glass plates. The pivot motor rotates the entire scanning assembly about the pivot point in order to align the scanning optics parallel to the gel between the glass plates. Reproduced from [7] with permission from VCH Verlagsgesellschaft mbH.*

required. The LI-COR instrument itself is also somewhat less expensive. Nevertheless, the higher throughput of the ABI 377 means that this is the preferred option for high-throughput laboratories or central sequencing facilities.

References

1. Richterich, P., Heller, C., Wurst, H. and Pohl, F.M. (1989) DNA sequencing with direct blotting electrophoresis and colorimetric detection. *BioTechniques*, **7**, 52–59.
2. Beck, S., O'Keeffe, T., Coull, J.M. and Koster, H. (1989) Chemiluminescent detection of DNA: application for DNA sequencing and hybridization. *Nucleic Acids Res.*, **17**, 5115–5123.
3. Tizard, R., Cate, R.L., Ramachandran, K.L., Wysk, M., Voyta, J.C., Murphy, O.J. and Bronstein, I. (1990) Imaging of DNA sequences with chemiluminescence. *Proc. Natl Acad. Sci. USA*, **87**, 4514–4518.
4. Church, G.M. and Kieffer-Higgins, S. (1988) Multiplex DNA sequencing. *Science*, **240**, 185–188.
5. Olesen, C.E., Martin, C.S. and Bronstein, I. (1993) Chemiluminescent DNA sequencing with multiplex labeling. *BioTechniques*, **15**, 480–485.

6. Ansorge, W., Sproat, B., Stegemann, J., Schwager, C. and Zenke, M. (1987) Automated DNA sequencing: ultrasensitive detection of fluorescent bands during electrophoresis. *Nucleic Acids Res.*, **15**, 4593–4602.

7. Middendorf, L.R., Bruce, J.C., Bruce, R.C. *et al.* (1992) Continuous, on-line DNA sequencing using a versatile infrared laser scanner/electrophoresis apparatus. *Electrophoresis*, **13**, 487–494.

8. Prober, J.M., Trainor, G.L., Dam, R.J., Hobbs, F.W., Robertson, C.W., Zagursky, R.J., Cocuzza, A.J., Jensen, M.A. and Baumeister, K. (1987) A system for rapid DNA sequencing with fluorescent chain-terminating dideoxynucleotides. *Science*, **238**, 336–341.

9. Lee, L.G., Connell, C.R., Woo, S.L., Cheng, R.D., McArdle, B.F., Fuller, C.W., Halloran, N.D. and Wilson, R.K. (1992) DNA sequencing with dye-labeled terminators and T7 DNA polymerase: effect of dyes and dNTPs on incorporation of dye-terminators and probability analysis of termination fragments. *Nucleic Acids Res.*, **20**, 2471–2483.

10. Tabor, S. and Richardson, C.C. (1995) A single residue in DNA polymerases of the *E. coli* DNA polymerase I family is critical for distinguishing between deoxy- and dideoxyribonucleotides. *Proc. Natl Acad. Sci. USA*, **92**, 6339–6343.

11. Reeve, M.A. and Fuller, C.W. (1995) A novel thermostable polymerase for DNA sequencing. *Nature*, **376**, 796–797.

12. Parker, L.T., Deng, Q., Zakeri, H., Carlson, C., Nickerson, D.A. and Kwok, P.-Y. (1995) Peak height variations in automated sequencing of PCR products using Taq dye-terminator chemistry. *BioTechniques*, **19**, 116–121.

13. Parker, L.T., Zakeri, H., Deng, Q., Spurgeon, S., Kwok, P.-Y. and Nickerson, D.A. (1996) AmpliTaq® DNA polymerase, FS dye-terminator sequencing: analysis of peak height patterns. *BioTechniques*, 21, 694–699.

14. Smith, L.M., Sanders, J.Z., Kaiser, R.J., Hughes, P., Dodd, C., Connell, C.R., Heiner, C., Kent, S.B. and Hood, L.E. (1986) Fluorescence detection in automated DNA sequence analysis. *Nature*, **321**, 674–679.

15. Qiagen (1995) The Qiagen Guide to Template Purification and DNA Sequencing. Qiagen GmbH, Hilden, Germany.

16. Ju, J., Glazer, A.N. and Mathies, R.A. (1996) Energy transfer primers: a new fluorescence labelling paradigm for DNA sequencing and analysis. *Nature Med.*, **2**, 246–249.

8 Troubleshooting

8.1 Introduction

DNA sequencing is a multi-step, complex procedure. It is normally highly reliable once established but, like any complex procedure, problems will occasionally arise.

The first question to ask is whether the problem is template-specific or systematic, by comparing different templates from the same gel. Template-specific problems affect certain templates only, while other reactions on the same gel are satisfactory. Systematic problems affect every reaction on the gel.

Template-specific problems are commonly caused by:

- using insufficient template DNA;
- using contaminated template DNA, for example contaminated with PEG, RNA; or other DNA;
- strong secondary structure in the template;
- no primer binding site in the template.

Systematic problems are commonly caused by:

- a defective reagent;
- suboptimal gel electrophoresis.

Using an inappropriate or incorrectly synthesized primer will appear as a template-specific problem if each reaction used a different primer, or as a systematic problem if every reaction on the gel used the same primer. Similarly, template problems can affect a whole gel if there is a problem with an entire batch of templates.

A more detailed analysis of common problems is presented below, categorized by the symptoms that they present (see *Table 8.1*). Two common problems are then discussed in more detail. This analysis is focused principally on manual sequencing, but many of the problems are shared by automated sequencers.

TABLE 8.1: *Troubleshooting*

Symptoms	Likely causes	Possible solutions
No bands visible in one or more sets of reactions	Insufficient or contaminated template	Check template on agarose gel Try another primer on same template Re-purify template
	No primer site in template	Check sequence of primer Try primer on another template
	Reagent omitted Defective reagent	Sequence control template Try sequencing control template Try fresh radioisotope (dNTPs are degraded by repeated freeze–thaw cycles Buy a new kit
Radioactivity remains at top of gel	Samples not denatured when loaded on gel	Just before loading gel, heat samples to 85–95°C for at least 2 min and then cool rapidly on ice
Bands not sharp and well-defined		
Bands smeary	Contaminated template DNA	Re-purify template
	Gel problem	Depends on gel system. Check gel temperature during run (40–50°C for Long Ranger, 50–55°C for acrylamide). Try using fresh acrylamide solutions. Wash plates carefully
Bands wavy 	Bubbles or dust in gel	Make another gel. Allow gel to polymerize completely. Check gel for bubbles and avoid these lanes. Place comb with care to avoid damaging wells
	Urea in wells	Use Pasteur pipet or syringe to wash out wells immediately before loading
Bands fuzzy	Poor contact between gel and film	Ensure gel is clamped firmly to gel. Avoid wrinkles when drying gel
Background smear	RNA contamination of template	Re-purify template

TABLE 8.1: Continued

Symptoms	Likely causes	Possible solutions
Some sequences faint	If entire gel affected, probably insufficient label or defective reagent	Use fresh label (dNTPs are degraded by repeated freeze–thaw cycles)
		Use more label
		Buy a new kit
Sequence faint in one lane	Component omitted or inadequately mixed	Repeat reactions
G A T C	ddNTP:dNTP ratio too low (or too high)	If using 'home-made' termination mixes, increase (or decrease) the amount of ddNTP
	Biased nucleotide usage usage in template, e.g. poly(A) tail	Add more of depleted dNTP, e.g. 1 μl of 0.5 M dTTP
Sequence faint near primer	ddNTP: dNTP ratio too low	If using 'home-made' termination mixes, increase the amount of ddNTP
	Insufficient annealed template	Use more of template DNA
		Check primer concentration
		See Section 8.2.3
Sequence fades away from primer	ddNTP:dNTP ratio too high	If using 'home-made' termination mixes, decrease the amount of ddNTP
Bands across more than one lane		
Bands across all four lanes	'Co-terminations'	See Section 8.2
Bands across two or three lanes	'Compressions'	See Section 8.3
G A T C	Mixed template or primer	Re-purify template from single colony or plaque. for M13 with large inserts, grow for only 5 h to minimize risk of spontaneous deletions
		Check primer on control template
Uneven band spacing	'Compressions'	See Section 8.3
Smile/frown on gel	Gel temperature uneven	Use electrophoresis apparatus with heat sink
	Gel of uneven thickness	When pouring gel, do not clamp gel plates together in regions not supported by spacers

8.2 Co-termination

Co-terminations, also known as 'pauses', 'false stops' or 'strong stops', are caused when the polymerase stops elongating the chain without incorporating a ddNTP. This will lead to a band at this point in the sequence independently of which ddNTP is present in the reaction, in other words a band in all four lanes (*Figures 8.1 and 10.8*). This will be superimposed on the 'true' band, so in a mild case it may be possible to assign the strongest band. In a severe case, this nucleotide will be completely impossible to read. In general, repeat with a fresh template preparation, making sure that the reaction components are thoroughly mixed (without introducing bubbles), and that the reaction temperatures are correct (see *Protocol 3.2*). Co-terminations can be completely overcome by using labeled ddNTPs, so that only those strands terminated with a ddNTP are detected (see Sections 4.4 and 7.3.1).

There are several possible reasons why the polymerase might fail to extend a chain even though it has not been terminated by incorporation of a ddNTP.

8.2.1 Secondary structure

The polymerase needs a single-stranded template. If a stable secondary structure is present, then the polymerase may not be able to read through it (see *Figure 3.5*). Secondary structure can be destabilized by heat or formamide. Co-termination due to secondary structure can, therefore, be overcome by increasing the temperature

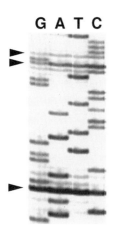

FIGURE 8.1: *Co-terminations lead to bands across all four lanes (arrowheads).*

of the sequencing reaction, or by adding formamide. Try using a thermostable polymerase, or adding formamide. Secondary structure can also be destabilized by the addition of 0.5 µg of single-stranded DNA-binding protein (Amersham/USB) to the reaction. This must be destroyed before loading on to the gel by incubation with proteinase K (0.1 µg proteinase K at 65°C for 20 min). Alternatively, sequence the problematic region on the other strand, as it is extremely unlikely that the same nucleotides will be affected.

8.2.2 Dirty template

The polymerase can be inhibited by contaminants in the template preparation. This can increase the frequency with which the polymerase dissociates from the template and so lead to an increase in co-terminations. This usually leads to a large number of co-terminations in the same reaction. The solution to this problem is to purify fresh template and repeat. Single-stranded templates seem to be less prone to this problem, so it is best to use a single-stranded template if possible.

8.2.3 Sequencing near to the primer

During the labeling reaction, the enzyme extends the primer, incorporating labeled (and unlabeled) nucleotides. These extended reaction products are then extended further during the termination reaction. Insufficient template DNA leads to longer extension products from the labeling reaction, so fewer labeled products are terminated by ddNTPs close to the primer. This results in faint sequence close to the primer. Inefficient extension in the termination reaction can lead to the products of the labeling reaction appearing as bands in all four lanes close to the primer. This has the characteristic appearance of a ladder of bands across all four lanes near the primer, fading after 100 or so bases and changing to an acceptable sequencing ladder further up the gel. Inefficient extension is typically due to poor quality template. Sequence near to the primer can be improved by the addition of Mn^{2+} to the labeling reaction (e.g. 1 µl of 100 mM $MnCl_2$). Manganese ions increase the frequency at which Sequenase incorporates ddNTPs, relative to dNTPs, and so shortens the average length of the terminated chains. Addition of Mn^{2+} allows sequence to be obtained from 20–200 nucleotides from the primer.

8.2.4 Incorrect dNTP concentration

Too low a concentration of one dNTP will lead to the polymerase pausing at this nucleotide and so may lead to co-terminations. Use of commercial dNTP mixes will avoid this problem in most cases, but a long homopolymeric run in the template may specifically deplete one nucleotide. A well-known example of this problem is the poly(A) tail of cDNAs. Sequencing through such a sequence will specifically deplete the reaction mix of the complementary nucleotide (dTTP), resulting in co-terminations and faint sequence in the T lane beyond the poly(A). This can often be overcome by increasing the concentration of the relevant nucleotide in the reaction mix, for example 1 µl of 0.5 M dTTP. In my experience, poly(A) tracts tend to make sequence faint or ambiguous for up to 20 nucleotides beyond the poly(A).

8.2.5 Reaction conditions

Co-terminations can be caused by inappropriate temperatures at various points in the reaction. In particular, performing the labeling reaction above 20°C may lead to co-terminations near the primer.

Removing the reaction from ice at the start of this reaction, as recommended in *Protocol 3.2* should avoid this problem. Similarly, the termination reactions should be performed at at least 37°C, so it is best to ensure that the termination mixes are pre-warmed for at least a minute. Termination reactions are best performed at 40–45°C, and can be run hotter if necessary. Glycerol can be added to stabilize the T7 DNA polymerase at elevated temperatures, but this requires modifications to the gel electrophoresis (see Section 6.4.2).

8.2.6 dITP

dITP is not incorporated as efficiently as dGTP, and so increases the level of co-termination. This effectively prohibits the routine use of dITP, though it is invaluable in resolving compressions (see below).

8.3 Compressions

Compressions are regions of the gel in which the normally even spacing of the bands is disrupted by anomalous migration of specific bands (see *Figure 8.2*). Extreme cases can even reverse the correct order of two bands. This is due to the effect of secondary structure on gel mobility. Since G–C base pairs are more stable than A–T pairs, GC-rich templates are particularly prone to this problem. Secondary structure problems are unlikely to affect the same bands when the same region is sequenced in both directions, so this is the most reliable method for detecting these anomalies. Compressions are probably the major cause of errors in published sequence and the reason for the absolute necessity of sequencing a template on both strands (i.e. in both directions). Strategies for resolving compressions all rely on reducing the degree of secondary structure formation in the gel.

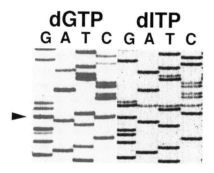

FIGURE 8.2: *Compressions are distortions of the band spacing. Here, sequencing with dGTP gives band spacing so distorted that the relative order of two bands cannot be determined (arrowhead). Re-sequencing the same template with dITP resolves the compression. Note that the dITP sequence contains a co-termination not seen in the dGTP reactions.*

8.3.1 Base analogs

Compressions can be resolved by replacing dGTP in the reaction mix with base analogs such as dITP, as I–C base pairs are less stable than G–C pairs (*Figure 4.5*). Unfortunately, dITP reactions cannot be read as far as those of dGTP, and are more prone to co-terminations (see Section 6.2) and other artifacts. dITP, therefore, is used only for resolving potential compressions, and not for routine sequencing. Reactions with labeled ddNTPs are an exception – co-terminations are completely eliminated and so dITP can be used routinely. Other base analogs such as 7-deaza-dGTP are available for the same purpose, but I find dITP to be completely effective in resolving compressions. The Sequenase® v2.0 kit (Amersham/USB) includes both dGTP and dITP mixes, and also pyrophosphatase to overcome pyrophosphorolysis, a slow degradation of the DNA which is a particular problem with dITP reactions.

8.3.2 Formamide gels

Formamide lowers the melting temperature of DNA and so destabilizes secondary structure. Gels can be made with formamide for this purpose, as discussed in Section 6.3.3.

Reference

1.   Tabor, S. and Richardson, C.C. (1989) Effect of manganese ions on the incorporation of dideoxynucleotides by bacteriophage T7 DNA polymerase and *Escherichia coli* DNA polymerase I. *Proc. Natl Acad. Sci. USA*, **86**, 4076–4080.

9 Confirmatory Sequencing

9.1 Introduction

The major use of DNA sequencing in many laboratories is to re-sequence regions of DNA whose sequence is already known or can be predicted. The sequence may be required to check that a recombinant DNA construct or PCR product has the expected sequence, or to determine the sequence of an allelic variant of a known sequence. Both are discussed below.

9.2 Checking constructs

Recombinant DNA technology allows the construction of novel DNA molecules, which may be formed from the ligation of two formerly noncontiguous DNA molecules, or by the modification of a pre-existing molecule by directed mutagenesis. It is often essential that the precise sequence of a part of the new molecule is determined to confirm that it is as intended. For example, the reading frame across the junction of two ligated molecules may be critical to the correct expression of a protein. Similarly, the sequence of a newly mutagenized DNA molecule must be confirmed by sequencing, as the methods available are not 100% reliable.

If the sequence of the rest of the molecule is known, it is quickest to synthesize a primer a short distance away from the sequence of interest, and to use this to sequence through this region. The primer should normally be 150–200 bases from the region of interest. This allows optimum resolution of the sequence of the region. Of course pre-existing primers can be used, as long as they are close enough to resolve the sequence reliably. In the case of a mutagenized molecule, it may be helpful to sequence the original, unmutagenized molecule in parallel, allowing a direct comparison between the two sequences.

9.3 Sequencing allelic variants

Geneticists today wish to know the molecular basis of the mutants that they study. This is often crucial evidence in the identification of a human disease gene – demonstrating that the candidate gene has molecular lesions in all or most of the patients analyzed. Sequence variation between mutant and wild-type variants may give important insights into the functional regions of the gene product. This is also the information required for medical diagnostic purposes (see Section 10.3).

Large-scale DNA rearrangements, insertions and repeat expansions can be detected by Southern blotting and hybridization to detect restriction fragment length polymorphisms (RFLPS), but point mutations generally cannot. Other molecular techniques such as single-strand conformational polymorphism (SSCP) [1, 2] and double-strand denaturing gel electrophoresis (DDGE) [3] can help to locate a point mutation, but only DNA sequencing can determine the molecular nature of the mutation precisely.

If the mutant allele has been cloned, then the sequence can be determined easily using synthetic oligonucleotide primers, basing the sequence of these primers on the known sequence of the wild-type gene. If the mutant allele has not already been cloned, the quickest method is to amplify the region of interest by PCR [or reverse transcriptase (RT)–PCR], and then to sequence the PCR reaction product. This is described in Chapter 11. In either case, of course, the wild-type sequence needs to be known for comparison with the mutants, and as the basis for primer design. If the wild-type sequence is to be determined alongside the mutant sequences, this would suggest that the 'primer walking' strategy may be the most appropriate (see Section 11.3).

It is often helpful to sequence a wild-type template alongside the mutants, even if the wild-type sequence is already known. This allows direct side-by-side comparison of the two sequences. Even if the data quality is too poor for unambiguous determination of the sequence, such a comparison often shows whether or not it is the same as the wild-type sequence.

9.4 Alternatives to DNA sequencing

There are several methods available for determining the DNA sequence of a very small region of DNA, typically a single nucleotide. These methods have a number of important applications, for example in clinical diagnosis of mutations responsible for genetic diseases. The sequence variation in the abnormal allele of many major disease genes has already been determined. It is therefore much quicker and easier to test for the presence of these known alleles, rather than to determine the entire sequence of both copies of the gene from each patient.

9.4.1 Using restriction endonucleases

Restriction endonucleases are valuable tools for checking constructs. Since the desired outcome of the manipulations is known precisely, it is often possible to identify newly created restriction sites that distinguish the correct product from likely alternatives. (see *Figure 9.1*). Characterizing a construct by restriction enzyme digestion is

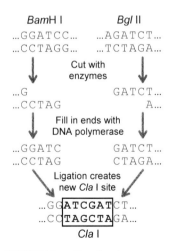

FIGURE 9.1: *Confirming sequence by use of restrictions endonucleases. In this example, an end-filled* BamHI *site is ligated to an end-filled* BglII *site. The predicted sequence contains a new* ClaI *site. The presence of this site can be confirmed by restriction enzyme digestion. Gain or loss of a single base in joining the two molecules would give a product which does not contain a* ClaI *site at the junction.*

much quicker and cheaper than DNA sequencing. The limitations of this technique are obvious – not all constructs will conveniently present a diagnostic restriction site.

The traditional markers for human mutations are RFLPs. RFLP analysis by Southern blotting takes up to a week to perform and has been largely replaced for diagnostic purposes by PCR-based methods. Here, the region containing the RFLP is amplified by PCR, digested by the appropriate restriction endonuclease and analyzed by agarose gel electrophoresis to reveal the presence or absence of the restriction site. A useful control is to include another, nonpolymorphic site for the same enzyme in the amplified region, as a control for the restriction digest. While RFLPs make good markers for genetic analysis, once the DNA sequence of a mutant allele is known it is clearly more appropriate for diagnostic purposes to analyze the presence or absence of the mutation directly, rather than looking at a linked marker. While deletions may well be directly detected by RFLPs, it is highly unlikely that a point mutation will conveniently create or destroy a restriction site, so other methods are required for this purpose.

Restriction endonucleases can also be used to probe the methylation state of DNA [4]. Most (98%) CpG dinucleotides in human DNA are methylated. The exceptions typically mark the 5′ ends of genes and so are of great interest to molecular geneticists. *Msp*I and *Hpa*II both recognize and cut at CCGG, but *Msp*I will cut $C^{Me}CGG$ whereas *Hpa*II will not. Sites cut by both enzymes are therefore unmethylated, whereas those cut by *Msp*I only are methylated. Only those CpG dinucleotides which form part of a convenient restriction site can be probed in this way; to examine all the C residues requires sequencing with bisulfite (Section 10.4).

9.4.2 Using oligonucleotide hybridization

The melting temperature (T_m) of an oligonucleotide hybridized to its complementary sequence is very sensitive to mismatches, particularly towards the middle of the oligonucleotide. This can be used to identify specific allelic variants by hybridizing (annealing) allele-specific oligonucleotides to the DNA of interest. Stringent conditions are used which will only allow hybridization of oligonucleotides perfectly complementary to the target. The oligonucleotide is labeled in some detectable way, for example by radiolabeling, or by conjugation to a fluorophore or hapten. Detection of the annealed oligonucleotide then indicates the presence of the specific allele which the oligonucleotide was designed to detect (*Figure 9.2*) [5].

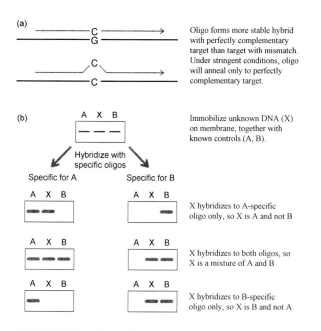

FIGURE 9.2:   *Allele detection by oligonucleotide hybridization. (a) A mismatch destabilizes a hybrid. (b) An unknown sample is immobilized on a membrane by 'Southern blot' or 'dot-blot'. Labeled oligonucleotides specific for each allelic variant 'A' and 'B' are then hybridized to this DNA. Hybrids are detected, e.g. by autoradiography. The pattern of hybridization indicates the composition of the unknown sample. This procedure can be reversed, by immobilizing the oligonucleotide and hybridizing with labeled sample.*

Using the method outlined above requires a separate hybridization for every oligonucleotide. This can be improved by immobilizing the oligonucleotide and probing with the gene of interest, usually as a labeled PCR product [6]. The only constraint now is that all the oligonucleotides have a similar T_m. This 'reverse dot-blot' method has the additional advantage of allowing direct comparison between various oligos and controls.

9.4.3 Using PCR

PCR depends on the use of oligonucleotide primers, which anneal to complementary DNA and prime DNA synthesis. PCR therefore lends itself to allele-specific detection by using primers specific for one allele rather than another. Of course this depends on the previous characterization of the various alleles by DNA sequencing. For PCR, the allele specificity is determined by the 3′ nucleotide of the primer, which complements one allele but not the other. *Taq* DNA polymerase lacks a 3′ → 5′ proofreading activity, and will not extend from a 3′ mismatch. Other thermostable polymerases such as *Pfu*, which do possess a 3′ → 5′ proofreading activity, are unsuitable for this application.

A typical assay comprises two PCR reactions using the same substrate DNA. The reactions use one primer in common and one allele-specific primer. Each reaction differs only in the use of a different allele-specific primer [7]. The products of the reaction are then analyzed by agarose gel electrophoresis. The presence of an amplification product in a reaction indicates the presence in the substrate of DNA of the allele to which the allele-specific primer in that reaction is fully complementary (*Figure 9.3*). The extreme sensitivity of PCR makes this type of analysis valuable for antenatal diagnosis.

A major advantage of using this PCR system to test for known allelic variants is that it easily can be multiplexed. Multiplexing is the simultaneous detection of allelic variants at multiple positions. Where the major allelic variants of a human disease gene have been characterized (by exhaustive DNA sequencing), a patient can be analyzed for the presence of each of these variants simultaneously [8–10]. This is clearly a valuable diagnostic tool. It is essential that the products of the multiplexed reactions can be distinguished from each other, but this is accomplished easily by choosing the common primers such that each multiplexed reaction amplifies a fragment of a characteristic size.

A further sophistication is to combine the two alternative allele-specific primers into a single reaction and determine which one (or both) can prime from the test sample [11]. Both reaction products will be the same size, so the oligonucleotide primers need to be labeled differentially. This can be arranged by labeling with fluorophores and detecting the label directly during gel electrophoresis using, for example, an ABI semi-automated sequencer [12]. This equipment also lends itself to multiplexing, so simultaneous assays for a number of allelic variants can be performed in a single PCR reaction, and analyzed on a single lane of the sequencer. Large numbers of samples can be analyzed rapidly using this approach.

PCR is notoriously sensitive to contamination and artifact, so it is imperative that an appropriate range of controls are present in every assay. In the allele detection experiments described above, appropriate controls might be to perform the assays in parallel on

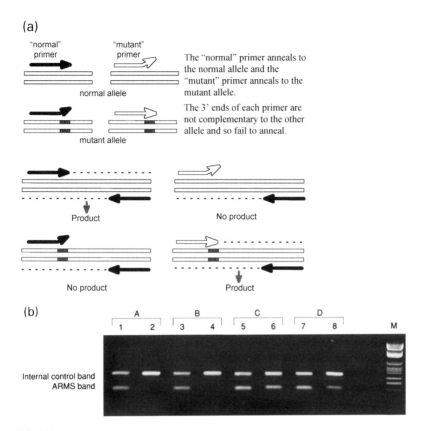

FIGURE 9.3: *Allele detection by PCR. (a) The 3' end of the primer must be complementary to the template to allow elongation and amplification. PCR with allele-specific primers differing at their 3' ends can therefore discriminate between different alleles. This method is variously known as ARMS (amplification refractory mutation system), ASP (allele-specific PCR), PASA (PCR amplification of specific alleles) and ASA (allele-specific amplification). (b) Analysis of β-hexosaminidase A exon 11. Odd-numbered lanes contain the products of the common and normal primers; even-numbered lanes contain the products of the common and mutant primers. Individuals A and B are normal for this allele, whereas individuals C and D are heterozygotes. Part (b) reproduced from Newton and Graham (1994) PCR, BIOS Scientific Publishers Ltd.*

known alleles. Multiplexing may introduce its own internal controls – no patient is likely to have more than one or two mutant variants of the gene examined, and so should be wild-type for most of the variants tested.

References

1. Orita, M., Iwahana, H., Kanazawa, H., Hayashi, K. and Sekiya, T. (1989) Detection of polymorphisms of human DNA by gel electrophoresis as single-strand conformation polymorphisms. *Proc. Natl Acad. Sci. USA*, **86**, 2766–2770.

2. Makino, R., Yazyu, H., Kishimoto, Y., Sekiya, T. and Hayashi, K. (1992) F-SSCP: fluorescence-based polymerase chain reaction single-strand conformation polymorphism (PCR-SSC) analysis. *PCR Methods and Applications*, **2**, 10–13.

3. Myers, R.M., Sheffield, V.C. and Cox, D.R. (1988) in *Genome Analysis: A Practical Approach* K. Davies, ed.). Oxford University Press, Oxford, pp. 95.

4. Nelson, M. and McClelland, M. (1992) Effect of site-specific methylation on DNA modification methyltransferases and restriction endonucleases. *Nucleic Acids Res.*, **20 Suppl.**, 2145–2157.

5. Saiki, R.K., Bugawan, T.L., Horn, G.T., Mullis, K.B. and Erlich, H.A. (1986) Analysis of enzymatically amplified beta-globin and HLA-DQα DNA with allele-specific oligonucleotides. *Nature*, **324**, 163–166.

6. Saiki, R.K., Walsh, P.S., Levenson, C.H. and Erlich, H.A. (1989) Genetic analysis of amplified DNA with immobilised sequence-specific oligonucleotide probes. *Proc. Natl Acad. Sci. USA*, **86**, 6230–6234.

7. Newton, C.R., Graham, A., Heptinstall, L.E., Powell, S.J., Summers, C., Kalsheker, N., Smith, J.C. and Markham, A.F. (1989) Analysis of any point mutation in DNA. The amplification refractory mutation system (ARMS) *Nucleic Acids Res.*, **17**, 2503–2516.

8. Chamberlain, J.S., Gibbs, R.A., Ranier, J.E., Nguyen, P.N. and Caskey, C.T. (1988) Deletion screening of the Duchenne muscular dystrophy locus via multiplex DNA amplification. *Nucleic Acids Res.*, **16**, 11141–11156.

9. Beggs, A.H., Koenig, M., Boyce, F.M. and Kunkel, L. (1990) Detection of 98% of DMD/BMD gene deletions by polymerase chain reaction. *Hum. Genet.*, **86**, 45–48.

10. Ferrie, R.M., Schwarz, M.J., Robertson, N.H., Vaudin, S., Super, M., Malone, G. and Little, S. (1992) Development, multiplexing, and application of ARMS tests for common mutations in the CFTR gene. *Am. J. Hum. Genet.*, **51**, 251–262.

11. Gibbs, R.A., Nguyen, P.N. and Caskey, C.T. (1989) Detection of single DNA base differences by competitive oligonucleotide priming. *Nucleic Acids Res.*, **17**, 2437–2448.

12. Chehab, F.F. and Kan, Y.W. (1990) Detection of sickle cell anaemia by colour DNA amplification. *Lancet*, **335**, 15–17.

10 Sequencing PCR Products

10.1 Introduction

PCR is an *in vitro* technique which allows the amplification of a specific DNA region that lies between two regions of known DNA sequence [1–3]. Oligonucleotides complementary to the regions of known sequence are synthesized and used as primers for DNA synthesis. In addition to the primers and the template DNA, the reaction components are simply a DNA polymerase, dNTPs and the various buffers and salts required for polymerase activity, notably Mg^{2+}. The method is illustrated in *Figure 10.1*. The exponential nature of the amplification means that specific DNA fragments can be amplified about a million-fold from tiny amounts of template DNA.

Since its introduction in 1985, PCR has become a major tool in molecular biology and new applications are constantly being developed. Technical advances include the introduction of thermostable DNA polymerases [4], of which the best known is *Taq* DNA polymerase, and the development of programable temperature control devices (known as thermal cyclers or PCR machines). The enormous value of the technique was recognized by the award of the 1993 Nobel Prize for Chemistry to Dr Kary Mullis.

PCR can be used as a method of template preparation by amplifying the region of interest from a single colony or plaque [5, 6]. This replaces the normal procedure of growing up the clone and purifying DNA from the culture. PCR is easily automated, so this may be preferred for high-throughput sequencing projects.

An unpurified PCR product is not normally a suitable template for DNA sequencing. This is because several components of the reaction are likely to interfere with the sequencing reaction, for example

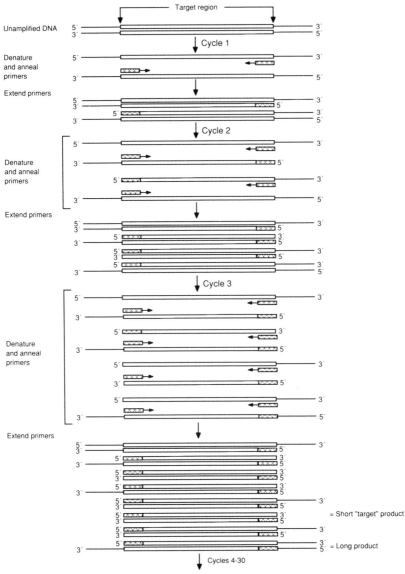

Amplification of short "target" product

FIGURE 10.1: *The polymerase chain reaction. PCR is a cycling process; with each cycle the number of DNA targets doubles. The strands in the targeted DNA are separated by thermal denaturation and then cooled to allow primers to anneal specifically to the target region. DNA polymerase is then used to extend the primers in the presence of the four dNTPs and suitable buffer. In this way duplicates of the original target region are produced. This 'cycle' is normally repeated 20–40 times. The short 'target' products increase exponentially after the fourth cycle whereas the long products increase only linearly. Reproduced from Newton and Graham (1994)* PCR, *BIOS Scientific Publishers Ltd.*

dNTPs and primers. Various methods are available for purifying the amplified DNA away from these contaminants, as described in Section 5.4. Detailed information about PCR can be found in another volume in this series [7].

10.2 Sequence information from PCR products

Several approaches are available for sequencing PCR products. Perhaps the most obvious is simply to subclone the PCR product into a plasmid or phagemid vector and then sequence it. It is essential, however, to realize that a purified PCR product is a pool of DNA molecules which are not entirely identical in sequence. This is because the thermostable DNA polymerases used have a noticeable error rate, and so will occasionally introduce errors [8]. These errors are then amplified in the subsequent rounds of replication.

Typically, a single molecule from a 1 kb PCR product synthesized using *Taq* DNA polymerase will have a small number (possibly zero) of single base changes relative to the original template. These errors arise essentially at random, so different molecules from the same reaction may or may not share these errors, while molecules from another PCR amplification using identical template, primers and conditions are unlikely to share the same errors (see *Figure 10.2*). Other thermostable polymerases are available with a lower, but still finite, error rate, for example *Pfu* DNA polymerase, which is claimed to have a 12-fold higher fidelity than *Taq*.

10.2.1 Sequence analysis of PCR products

The imperfect fidelity of PCR has important implications for the sequencing of PCR products. Subcloning a PCR product effectively selects one molecule from the mix and clonally amplifies it. If the purpose of the PCR is to produce DNA for subcloning, perhaps introducing convenient restriction sites, or amplifying from a hard-to-obtain template, then it is important to sequence the subclone. This confirmatory sequencing ensures that the subclone has the expected sequence, without any unwanted mutations. Conversely, if the purpose of the sequence determination is to deduce the sequence of the template DNA used in the original PCR, it is important not to be misled by polymerase errors during the amplification. There are two ways to overcome this problem.

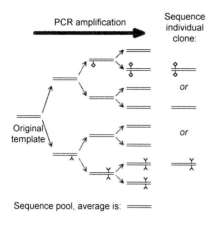

FIGURE 10.2: *PCR errors. DNA replication during PCR has a significant error rate. Errors are propagated through subsequent rounds of replication, so that the final product is a mixture of molecules, many of which have single base sequence changes relative to the original PCR template. Sequencing the pool identified the predominant base at each position. Cloning isolates a single molecule from the pool. This molecule may not perfectly represent the original template.*

Sequence independent subclones. Since independent PCR amplifications are unlikely to introduce the same error, one approach is to subclone an amplification product from each of three independent amplifications and sequence each of them. Sequence variation between the subclones can then be attributed to polymerase error. Since these errors are rare, any given error should appear in only one of the three subclones, and so the sequence of the original PCR template can be deduced.

Sequence the PCR product directly. The preferred approach is to sequence the PCR product directly. Unless an error was introduced in the first round of amplification from a very low copy-number template, any given error is rare in the pool of reaction products. The correct base will therefore be present in the overwhelming majority of molecules. If the sequence of many molecules is determined simultaneously, then the *average* of the sequences at any position will be a faithful reflection of the sequence of the template for the original PCR reaction (see *Figure 10.2*).

10.2.2 Fidelity of other polymerases

The problems associated with imperfect polymerase fidelity discussed here also apply in other contexts. While the 20 or so cycles of DNA synthesis in a PCR reaction exacerbate the problem, even single-pass reactions may introduce errors. For example, reverse transcriptases have relatively poor fidelities, so one must be careful in attributing variation between cDNAs to polymorphism in the genes from which these sequences were derived, rather than to errors introduced during reverse transcription. Similarly, end-filling by Klenow or T4 DNA polymerase, followed by blunt end ligation, is known to result in the gain or loss of a base at an appreciable frequency, necessitating confirmatory sequencing or other probing of the junction sequence to ensure that it is as intended.

10.3 Mutant detection by sequencing PCR products

Geneticists wish to know the molecular basis of the mutants they study, and clinicians require a reliable test for mutants leading to human genetic disease. It is usually far too time-consuming to clone the mutant gene by traditional methods of library construction from mutant tissue. In any case there may be only tiny mounts of DNA available, for example in antenatal screening. The answer is to amplify the DNA of interest by PCR. This may mean several overlapping amplifications, as the optimum size for PCR and sequencing is only a few hundred bases, compared with a typical gene size of at least a few kilobases.

Detection of a homozygous recessive mutation is relatively straightforward, as all of the amplified DNA will contain the mutant sequence. Sequencing the pool of PCR products will therefore show a different sequence at the site of the mutation, relative to the wild-type sequence. It is often helpful to run a wild-type control alongside the experimental samples for direct comparison. Quantifying the ratio of allelic forms is more problematical. In germ-line samples from heterozygotes it will be 50:50, but any ratio is possible in mixed viral populations, messenger RNAs (mRNAs) or modified DNAs. Detecting and quantifying the levels of different alleles then requires careful normalization of signal strength, and careful optimization of the PCR conditions to minimize differential amplification of specific variants [9]. Detection of 5-methylcytosine (5-MeC) residues is discussed below as an example of such a method (see Section 10.4).

More straightforward is the detection of dominant mutations, or of carriers of recessive mutations or polymorphisms. This is because the individual has two wild-type copies of the gene, two mutant copies or one of each. There are, therefore, only three possible outcomes to be distinguished. The PCR reaction will amplify both of these, and sequencing will therefore give a mixture of wild-type and mutant sequence. In the case of point mutations, this means that the site of the mutation will appear as a mixture of the wild-type and the mutant bases (*Figure 10.3*).

Diagnostic laboratories require high throughput, with rapid and accurate identification of any mutations. This is an ideal situation for automated sequencing (see Chapter 7). Unfortunately, the peak height variations generated by dye terminator chemistry (Section 7.3.1) make heterozygote detection difficult. With dye primers (Section 7.3.2), the interpretation is more straightforward –

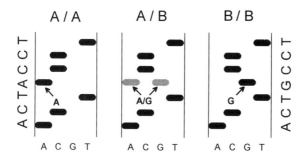

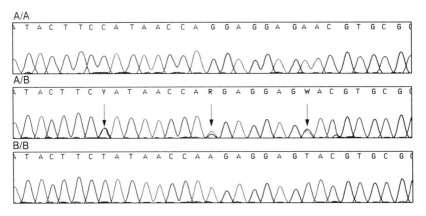

FIGURE 10.3: *Heterozygote detection. A heterozygote is detected as a mixture of two bases at one position. Comparison with control homozygous templates allows the confirmation of the presence of a heterozygote, rather than a sequencing artifact. Automated sequence data are from the HLA-DRB1 gene, obtained using dye primer chemistry. Reproduced from [10] with permission from Perkin-Elmer Applied Biosystems Division.*

homozygotes have peaks of uniform height, whereas a heterozygote will have a double peak at the site of the mutation, each peak being half the height of a homozygous peak. Dye primer sequencing does, however, require a set of dye-labeled primers capable of annealing to the sequencing template, in this case a PCR product. This means either incorporating the sequence of a standard dye primer into the PCR product, or else synthesizing a custom set of dye primers.

10.3.1 Tailed primers

In this method, one PCR primer has the −21M13 sequence added at its 5′ end, the other has the M13 Rev primer sequence similarly added. The PCR product is then suitable for sequencing with standard dye primer sets: −21M13 to sequence from one end and M13 Rev to sequence from the other (*Figure 10.4*). The sequences of these two primers are:

Primer	Sequence
−21M13	5′-TGTAAAACGACGGCCAGT
M13 Rev	5′-CAGGAAACAGCTATGACC

Advantages. The main advantage of this method is that both strands can be sequenced from a single PCR reaction. Furthermore, the use of only two sets of sequencing primers allows reaction conditions to be optimized, standardized and potentially automated.

Disadvantages. The 18-base tails shown above must be added to the PCR primers when they are synthesized. This increases the cost of the primers, particularly since long primers may need purification prior to use.

10.3.2 Custom dye primers

In this method, the PCR primers are neither tailed nor derivatized. The PCR product is sequenced using custom dye primers (*Figure 10.5*). These should be designed to be complementary to an internal region of the PCR product, so that primer artifacts from the PCR reaction do not affect the sequencing reaction. This method allows a single long PCR product to be sequenced using several sets of dye primers, minimizing the number of PCR reactions that need to be performed.

Advantages. PCR primers are cheaper, existing oligonucleotides can be used and long PCR products can be sequenced using different sets

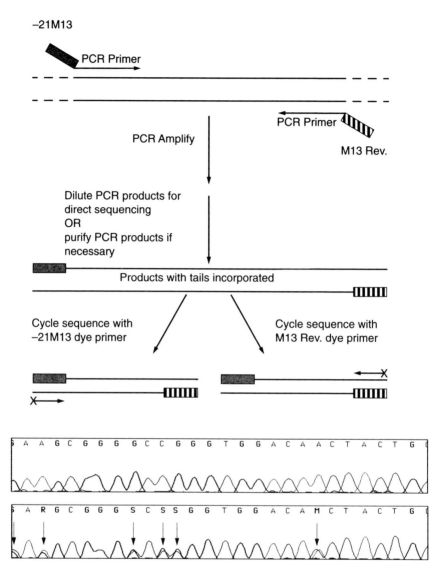

FIGURE 10.4: *Use of tailed PCR primers and typical sequence data. Reproduced from [10], with permission from Perkin-Elmer Applied Biosystems Division.*

of primers. This method is more tolerant of primer artifacts in the template, as the sequencing primers are not complementary to the PCR primers.

Disadvantages. Custom dye primers are required, which are much more expensive than standard primers. If the region is to be sequenced repeatedly, the additional cost per sample may be

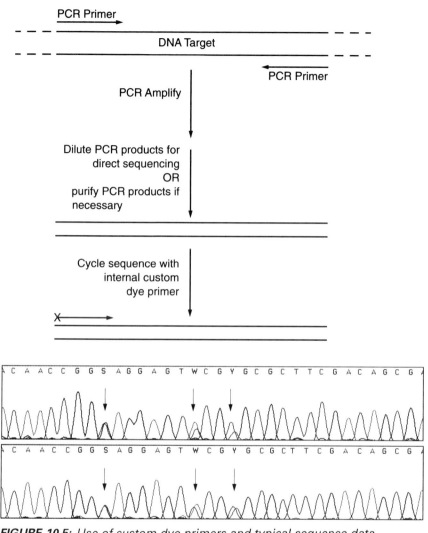

FIGURE 10.5: *Use of custom dye primers and typical sequence data. Reproduced from [10], with permission from Perkin-Elmer Applied Biosystems Division.*

acceptable. Custom primers should be tailed with the first five bases of the M13 Rev primer (5'-CAGGA-3') to allow correct software compensation for mobility effects visible on fragments up to 50 or so bases long (see Section 7.3.2).

10.3.3 Dye terminators

Despite the variable peak height, heterozygotes can be detected using dye terminator chemistry (*Figure 10.6*). However, for reliable results, this requires that both strands be sequenced, and that each apparent deviation from wild-type is checked carefully on both strands. This is clearly more labour intensive than dye primer methods, which need much less manual scrutiny. On the other hand, dye terminator sequencing needs no tailed or derivatized primers and so is likely to be considerably cheaper.

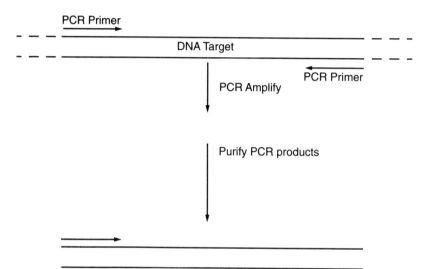

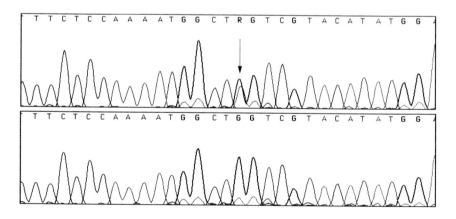

FIGURE 10.6: *Heterozygote detection with dye terminators. Reproduced from [10] with permission from Perkin-Elmer Applied Biosystems Division.*

10.3.4 Confirming the presence of heterozygotes

Compressions, co-terminations and other sequence artifacts can lead to false-positive or false-negative identification of heterozygotes. In each case, the most reliable method for confirming the heterozygote is to sequence the region in the other direction. It is extremely unlikely that a sequence-specific artifact will affect the same nucleotide in this case, as the sequence context of the second strand is unrelated to the original sequence (*Figures 10.7* and *10.8*).

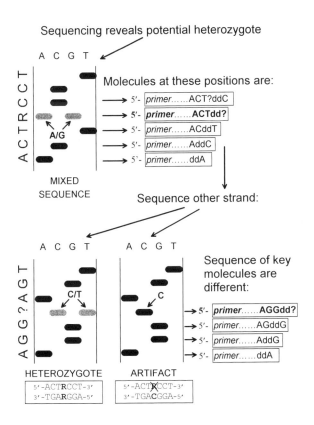

FIGURE 10.7: *Confirming the presence of a heterozygote. A mixture of two bases is detected at one position. This could be due to a mixed template (e.g. a heterozygote) or to a sequencing artifact caused by secondary structure. Sequencing the region on the other strand generates a set of molecules unrelated in structure. Detection of mixed bases confirms the presence of a heterozygote, failure to detect mixed bases implies that the original detection was an error due to secondary structure.*

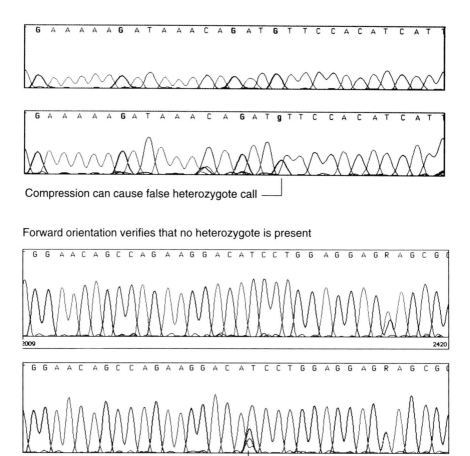

Compression can cause false heterozygote call

Forward orientation verifies that no heterozygote is present

A stop can cause a false positive

FIGURE 10.8: *Compressions or co-terminations can appear as heterozygotes. Dye primer chemistry is relatively prone to compressions and co-terminations, either of which can be confused with heterozygotes. Sequencing the other strand will normally resolve this. Reproduced from [10], with permission from Perkin-Elmer Applied Biosystems Division.*

10.4 Sequencing methylated DNA

Bisulfite treatment provides an efficient method for detecting individual 5-MeC residues in DNA [11, 12]. Sodium bisulfite

A C T G^(me)C T G A^(me)C T G C^(me)C A   DNA sample

    ↓ Treat with NaHSO₃

A C T G **U** T G A **U** T G C **U** A

    ↓ Amplify by PCR

A C T G **T** T G A **T** T G C **T** A

    ↓ Sequence, compare with untreated

A C T G **T** T G A **T** T G C **T** A   NaHSO₃
A C T G **C** T G A **C** T G C **C** A   untreated
A C T G^(me)**C** T G A^(me)**C** T G C^(me)**C** A   deduced sequence

FIGURE 10.9: *Detecting 5-methylcytosine. DNA to be analyzed is treated with sodium bisulfite using conditions which deaminate cytosine but not 5-MeC. The modified DNA is amplified by PCR. PCR replicates U as if it were T and 5-MeC as C. This amplified DNA is then sequenced. The sequence is compared with that of an identical sample which has not been treated with sodium bisulfite, and the pattern of methylation in the original DNA sample is deduced.*

deaminates cytosine residues in single-stranded DNA to uracil, under conditions where 5-MeC residues are not affected. Following treatment with sodium bisulfite, the DNA can be amplified by PCR. During this process, U and T residues are both amplified as T and only 5-MeC residues are amplified as cytosine (C). The sequence of the PCR product can be compared with that of control PCR product from untreated DNA (see *Figure 10.9*). The level of methylation at every cytosine can be determined by sequencing the PCR product directly, whereas the pattern of methylation in individual molecules can be determined by subcloning and sequencing individual molecules from the PCR product. Subcloning and sequencing is relatively straightforward – each residue is either a C or a T – but quantifying the C:T ratio in the PCR product accurately by direct sequencing is more problematic. A recently published method [12] uses a biotinylated and a –21M13 tailed primer in the initial PCR reaction to allow solid phase capture of the product on Dynal magnetic beads (Dynal AS, Oslo, Norway) followed by direct sequencing with –21M13 dye primers (see Sections 5.4.1 and 10.3.1). The same fluorochrome (FAM) was used for both C and T (or A and G) reactions, to allow direct comparison of peak heights. Sequence analysis software normalizes for different loadings in each reaction, and for the different properties of the four dyes used, so ABI GENESCAN analysis was used instead, which does not make such corrections.

References

1. Saiki, R.K., Scharf, S., Faloona, F., Mullis, K.B., Horn, G.T., Erlich, H.A. and Arnheim, N. (1985) Enzymatic amplification of beta-globin genomic sequences and restriction site analysis for diagnosis of sickle cell anemia. *Science*, **230**, 1350–1354.
2. Mullis, K.B., Faloona, F., Scharf, S., Saiki, R., Horn, G. and Erlich, H. (1986) Specific enzymatic amplification of DNA *in vitro*: the polymerase chain reaction. *Cold Spring Harbor Symp. Quant. Biol.*, **51**, 263–273.
3. Mullis, K.B. and Faloona, F.A. (1987) Specific synthesis of DNA in vitro via a polymerase-catalysed chain reaction. *Methods Enzymol.*, **155**, 335–350.
4. Saiki, R.K., Gelfand, D.H., Stoffel, S., Scharf, S.J., Higuchi, R., Horn, G.T., Mullis, K.B. and Erlich, H.A. (1988) Primer-directed enzymatic amplification of DNA with a thermostable DNA polymerase. *Science*, **239**, 487–491.
5. Walsh, P.S., Metzger, D.A. and Higuchi, R. (1991) Chelex 100 as a medium for simple extraction of DNA for PCR-based typing from forensic material. *BioTechniques*, **10**, 506–513.
6. Krishnan, B.R., Blakesley, R.W. and Berg, D.E. (1991) Linear amplification DNA sequencing directly from single phage plaques and bacterial colonies. *Nucleic Acids Res.*, **19**, 1153.
7. Newton, C.R. and Graham, A. (1997) *PCR*, 2nd Edn. BIOS Scientific Publishers, Oxford, UK.
8. Tindall, K.R. and Kunkel, T.A. (1988) Fidelity of DNA synthesis by the *Thermus aquaticus* DNA polymerase. *Biochemistry*, **27**, 6008–6013.
9. Larder, B.A., Kohli, A., Kellam, P., Kemp, S.D., Kronick, M. and Henfrey, R.D. (1993) Quantitative detection of HIV-1 drug resistance mutations by automated DNA sequencing. *Nature*, **365**, 671–673.
10. Perkin-Elmer/ABI (1995) *Comparative PCR Sequencing*. Perkin-Elmer/ABI, CA.
11. Grigg, G.W. and Clark, S.J. (1995) Sequencing 5-methylcytosine residues in genomic DNA. *BioEssays*, **16**, 431–436.
12. Paul, C.L. and Clark, S.J. (1996) Cytosine methylation: quantitation by automated genomic sequencing and GENESCAN™ analysis. *BioTechniques*, **21**, 126–133.

11 Strategies for New Sequence Determination

11.1 Introduction

Sequencing an unknown sequence by the chain termination method requires that the unknown sequence is adjacent to a short stretch of known sequence. Sequencing primers complementary to the known sequence are then used, allowing the sequence of the adjacent few hundred bases to be determined. However, *de novo* sequencing projects often require far more than a few hundred bases of new sequence. For example, cDNAs are typically a few thousand bases long and cosmids a few tens of kilobases. At the extreme, the human genome sequencing project requires the accurate determination of the entire human genome of approximately 3×10^9 bases. This chapter discusses the various methods and strategies whereby the sequence of large segments of DNA can be deduced, despite the limited range of individual sequence reactions.

Genomic sequencing projects will have an enormous impact on this type of work. The human genome and the genomes of most major biological models will be sequenced over the next few years or decades. Sequencing clones from these organisms will then become a matter of confirmatory rather than *de novo* sequencing, looking for mutations, variations, splice sites and so on. However, today's researchers do not want to wait a few years for a genome project to sequence their gene!

11.2 Directed versus nondirected strategies

Somehow each part of the large unknown DNA has to be brought close enough to known sequence to allow the unknown DNA to be sequenced. The strategies for doing this can be broadly categorized as directed or nondirected. Nondirected strategies generate sequence from different parts of the unknown DNA at random. If enough such sequences are obtained, then eventually the sequence of the whole molecule will have been determined. The set of sequences are compiled into a contiguous sequence ('contig') using a computer program which looks for overlaps between the sequences (see Chapter 15). In a directed strategy, the unknown DNA is sequenced in an orderly way, so that each sequencing reaction will come from a predictable part of the unknown DNA. This means that the individual sequences can be compiled into a contiguous sequence by hand, though in practice computers are always used (see Chapter 15). Specific strategies, both directed and non-directed, are summarized in *Table 11.1* and discussed in detail below.

TABLE 11.1: *Summary of strategies for* de novo *sequencing*

Method	Advantages	Disadvantages
Primer walking	Simple Minimal 'hands-on' time Produces set of primers spanning sequence	Slow Expensive Difficult to automate

Restriction endonuclease digestion and subcloning	Contig assembly relatively
	  insensitive to repeats
	Early production of
	  restriction map
		Considerable 'hands-on' time
		Difficult to automate
		Relies on presence of
		  convenient restriction sites

Shotgun sequencing	Cheap
	Quick
	Easily automated
		Becomes inefficient towards
		  end of sequencing project
		  as many regions are
		  sequenced repeatedly

Deletion series	Contig assembly relatively
	  insensitive to repeats
	Deletions may be useful
	  for functional analysis
		Deletion series may be hard
		  to generate

Transposon-facilitated sequencing	Cheap Quick Easily automated	Requires special strains Insert analysis limited by range of PCR

11.3 Primer walking

The simplest strategy is known as 'primer walking' (*Figure 11.1*). In this approach, a new sequencing primer is designed, based on the most distant reliable sequence obtained from the previous sequencing reaction. This primer is then used to sequence the next unknown section of the template. In this way, the sequence is determined in a methodical, stepwise fashion, starting from the ends and working towards the middle.

This method is relatively slow, as the sequence is only determined from the ends. Each 'step' requires a sequencing reaction, gel electrophoresis, data analysis, the design of a new primer and synthesis of the primer. This process will take at least several days, probably more if the primer is ordered from an external supplier. This time scale might be acceptable for sequencing a short cDNA, but not for a long cDNA, let alone a cosmid. This process is also difficult to automate – every reaction requires a different primer, and these cannot be made at the same time as each depends on the results of the previous reaction.

Primer walking is also relatively expensive. Each step requires the synthesis of a new sequencing primer. Until recently, cost

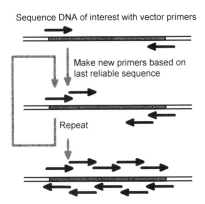

Sequence DNA of interest with vector primers

Make new primers based on last reliable sequence

Repeat

FIGURE 11.1: Primer walking. The ends of the DNA of interest are sequenced using vector-specific primers, e.g. 'universal' primers. Custom primers are then synthesized based on the new sequence information. This process is repeated until the entire region of interest has been sequenced.

considerations made this strategy impractical for all but the best-funded laboratories, but the costs of oligonucleotide synthesis have dropped markedly over the past few years, so this is no longer a major objection.

Primer walking does have several advantages. It is conceptually and technically simple, requiring no subcloning or other manipulations that might lead to rearranged clones or other problems. It therefore requires little 'hands-on' time, so that while the sequence is only determined slowly, the scientist has plenty of time to get on with other things.

Primer walking also generates a set of sequencing primers covering the entire sequence. If the sequence has to be determined repeatedly, for example to sequence mutant alleles, this set of primers becomes a very useful resource.

Since a single large template is being sequenced repeatedly, it is worth taking the trouble to ensure that this template is of optimum quality, so that the 'step' size is as large as possible, and errors and ambiguities in the sequence are minimized. With a phagemid template, purification of one strand by use of a helper phage (*Protocols 5.1 and 5.2*) and the other by use of NBL's Quick-Strand kit (*Figure 5.2*) allows all the sequencing to use single-stranded template.

11.4 Restriction endonuclease digestion and subcloning

Sequence information can, at least in principle, be obtained from any known restriction endonuclease site. Simply digest with the restriction enzyme and subclone an appropriate fragment, such that the unknown sequence near the restriction site is now adjacent to known sequence from the vector. This allows a primer complementary to the vector to be used to determine the unknown sequence. It is often convenient to delete a fragment of the unknown clone by cutting the clone at two (or more) sites and re-circularizing the digestion products with DNA ligase. As the enzymes chosen are unlikely to generate compatible cohesive ends, it will normally be necessary to convert these to blunt ends with Klenow or T4 DNA polymerase. This approach is illustrated in *Figure 11.2*.

The key requirement for this approach is an accurate restriction map with convenient restriction enzyme sites spaced every few hundred

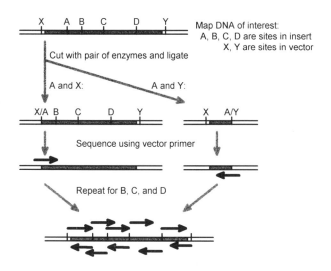

FIGURE 11.2: *Restriction mapping and subcloning. A detailed, accurate restriction map of the DNA of interest is prepared. This is used to generate deletions which bring different parts of the DNA of interest adjacent to known vector sequence. Alternatively, fragments may be subcloned into another vector. The subclones are sequenced using vector-specific primers, e.g. 'universal' primers. Note that the final sequence is usually incomplete, as shown here, with sections which have only been sequenced in one direction. Gaps can be filled using custom primers.*

bases. A restriction map can be generated very quickly by an experienced worker, but the distribution of restriction sites is a matter of chance, so it is not likely that there will be sufficient well-spaced sites to sequence a large clone completely using this method exclusively. However, it is generally possible to obtain a large proportion of the sequence by this method. The sequence obtained is based on the restriction map, so it is clear where the gaps in the sequence are and how big they are, which helps in planning the next step.

This approach is difficult to automate, as it relies on a set of 'custom' subcloning steps, which will be different for each sequencing project. Perhaps the most common use is to use only a small number of restriction sites spread across a large unknown clone. Each of these sites becomes an entry point into the unknown sequence, from which primer walking can be performed in each direction. This hybrid approach can dramatically reduce the time taken to complete a sequencing project relative to primer walking alone.

11.5 'Shotgun' methods

A large piece of DNA can be broken up randomly into many smaller fragments. These smaller fragments can then be subcloned, their ends sequenced, and the resulting sequence information compiled by computer to reassemble the sequence of the large piece of DNA [1]. This is known as 'shotgun sequencing' (*Figure 11.3*). This approach rapidly yields 90% or so of the desired sequence information, but then becomes less efficient as each subsequent random sequence fragment is more and more likely merely to repeat information already obtained. It is therefore generally quicker to sequence the last few unknown regions by synthesizing custom oligonucleotide primers [2].

Three methods are available to fragment a large piece of DNA: digestion with restriction enzymes; sonication; and digestion with DNase I in the presence of Mn^{2+}. For all of these methods, it is important first to purify the target DNA, away from its vector DNA, or a proportion of the fragments generated will contain only vector DNA.

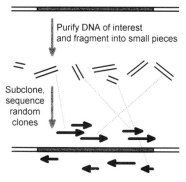

Purify DNA of interest
and fragment into small pieces

Subclone,
sequence
random
clones

Use computer to order the sequences,
based on overlaps.

FIGURE 11.3: *'Shotgun' sequencing. The DNA of interest is fragmented with enzymes or shear force. These random fragments are subcloned, then sequenced using vector-specific primers, e.g. 'universal' primers. The sequence information is compiled by computer. This method generates most of the sequence very quickly, but usually with some gaps, as shown. Custom primers are then usually synthesized to sequence across these gaps.*

11.5.1 Frequently cutting restriction endonucleases

Digestion with frequently cutting restriction enzymes such as *Sau*3A or *Taq*I gives fragments which are easy to subclone by virtue of their cohesive ends. Sequencing these subclones can give artifacts where two independent restriction fragments have been ligated together. It is important, therefore, to treat each such fragment separately when assembling the sequence data into a contig (see Chapter 15). The junction can easily be recognized as it will have the recognition sequence for the restriction enzyme used (e.g. GATC for *Sau*3A). Furthermore, the fragments from a single enzyme digestion do not overlap, and so cannot be assembled into a contig. It is therefore necessary to use at least two enzymes independently to generate a set of overlapping sequences. The main disadvantage of this approach is that the fragments generated are not truly random, but depend on the distribution of sites for the restriction enzyme(s) used. Small fragments are subcloned more efficiently than larger ones, so the small fragments tend to be sequenced many times over while the larger ones may not be sequenced at all.

11.5.2 Sonication

Large DNA molecules are very sensitive to shear force. Sonication can be used to shear DNA into fragments of 500–1000 bp, the ideal size for random sequencing [3]. This is the method of choice for shotgun sequencing as it is efficient and genuinely random (but see below).

11.5.3 DNase I digestion

In the presence of Mg^{2+}, pancreatic DNase I nicks (cuts one strand of) double-stranded DNA, whereas in the presence of Mn^{2+} it cuts both strands, generating a double-strand break. This activity can be used to derive random fragments in just the same way as sonication, but is a much less efficient method, and harder to control.

11.6 Transposon-facilitated sequencing

The approaches described above involve the juxtaposition of unknown sequence and known sequence (complementary to the primer) by moving the unknown sequence next to a known vector sequence. An alternative approach is to insert a known sequence into the unknown

DNA and sequence out from there [4, 5]. This can be achieved by using a transposon which integrates essentially at random into target DNA, for example the γδ transposon (Tn*1000*). This method is outlined in *Figure 11.4*. The sequence of the transposon is known, and may be engineered to have binding sites for standard primers. Although the initial integration is random, the set of clones generated can be sorted into a series by restriction mapping, or more rapidly, by PCR using primers specific to the vector and to the transposon. The details of this method, and the bacterial genetics involved, are beyond the scope of this book but are elegantly described in Strathmann *et al.* [6].

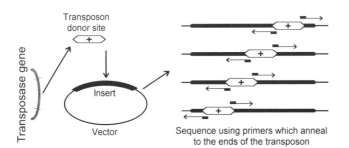

FIGURE 11.4: *Transposon-facilitated sequencing. The transposon has been engineered to carry a selectable marker (+). This allows the selection of those plasmids into which the transposon has jumped. The insertion site is then determined by PCR or restriction mapping, and selected clones are sequenced.*

11.7 Deletion series

The purpose of making a deletion series for sequencing is to bring progressively more distant regions of unknown DNA adjacent to the primer-binding site of the vector by progressively deleting all the DNA in between. The best method for constructing such a deletion series is to use exonuclease III (ExoIII), which digests DNA in a progressive and predictable way [7]. A major advantage is that vector DNA can be protected from the exonuclease, and so no additional subcloning is required after exonuclease digestion. This approach is outlined in *Figure 11.5*.

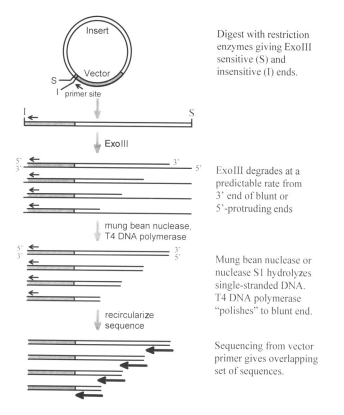

FIGURE 11.5: *Use of exonuclease III (ExoIII) to generate a deletion series.*

ExoIII removes nucleotides from the 3′-OH end of double-stranded DNA, but only from recessed or blunt 3′ ends and not from protruding ones (see *Table 11.2*). This allows the vector DNA to be protected by cutting it with two enzymes, one of which gives exonuclease-sensitive ends and the other exonuclease-resistant ends (e.g. *Kpn*I or *Sac*I), with the resistant site between the primer site and the unknown sequence (*Figure 11.5*). Note that restriction enzymes do not always cut two adjacent sites efficiently (see Section 11.7.3). The DNA is then digested with an excess of ExoIII, which degrades one of the two strands. The other strand is then removed with a single strand-specific nuclease such as mung bean nuclease. Mung bean nuclease does not give a high proportion of ligatable blunt ends, so the ends are treated further with the Klenow fragment of DNA polymerase I, or T4 DNA polymerase, either of which will fill in recessed 5′ ends and so leave clonable blunt ends. The shortened molecules can then be re-circularized using T4 DNA ligase.

TABLE 11.2: Blunt and protruding ends

Type of end	Enzyme	Recognition site			Cut ends		
5′ Protruding	BamHI	5′	GGATCC	3′	5′	G	3′
		3′	CCTAGG	5′	3′	CCTAG	5′
	EcoRI	5′	GAATTC	3′	5′	G	3′
		3′	CTTAGG	5′	3′	CTTAA	5′
Blunt	EcoRV	5′	GATATC	3′	5′	GAT	3′
		3′	CTATAG	5′	3′	CTA	5′
	SmaI	5′	CCCGGG	3′	5′	CCC	3′
		3′	GGGCCC	5′	3′	GGG	5′
3′ Protruding	KpnI	5′	GGTACC	3′	5′	GGTAC	3′
		3′	CCATGG	5′	3′	C	5′
	SacI	5′	GAGCTC	3′	5′	GAGCT	3′
		3′	CTCGAG	5′	3′	C	5′

Blunt and 5′ protruding ends are sensitive to ExoIII. 3′ Protruding ends are insensitive. 5′ Protruding ends can be protected against ExoIII by end-filling with thionucleotides.

Under these conditions, ExoIII digests all the DNA molecules in the reaction at essentially the same rate. This means that a single deletion reaction contains a population of molecules each with a similar amount of DNA deleted. A deletion series can, therefore, easily be constructed by removing aliquots from the exonuclease reaction at a series of time points.

ExoIII digestion is the best general method for generating a deletion series, but there are a few pitfalls not immediately apparent from the outline method above. The most frequently encountered difficulties are discussed below.

There are a number of commercially available kits for generating deletion series using ExoIII, but reliable and cheaper published protocols are also available [7–9].

11.7.1 Exonuclease digests too fast or too slow

While the exonuclease digests its substrate DNA at a uniform rate, this rate seems to vary from one DNA substrate preparation to another. This presumably reflects impurities in the DNA preparation. This problem can be circumvented by preparing an excess of DNA substrate, then performing the exonuclease digestion using a wider range of time points than is calculated to be necessary. If the enzyme is digesting at close to the nominal rate, the desired deletions will be within the first reaction set. If not, the reaction can be repeated using some of the remaining substrate preparation, with time points selected on the actual rate of exonuclease digestion, as measured in the previous experiment.

If the endonuclease does not appear to digest at all, this could mean that the enzyme is defective, but it is also possible that the substrate DNA has been prepared with no sensitive ends, as a consequence of failure of one restriction enzyme (see below).

11.7.2 DNA is completely degraded by exonuclease

ExoIII can digest from nicks (single-strand breaks) in the substrate DNA, as well as from the ends generated by appropriate restriction endonucleases. It is therefore important that the substrate DNA is largely free of nicks. This is true of DNA purified on a CsCl gradient, but may not be true of DNA purified by other methods, for example using a commercial kit. Most kits will produce perfectly satisfactory DNA if used carefully but, unlike CsCl purification, they do not actually separate nicked from supercoiled DNA. The quality of the DNA can be assessed easily by analyzing a small aliquot of undigested DNA by agarose gel electrophoresis – the large majority of the DNA should be supercoiled (higher mobility than the linear form), with only a small proportion of nicked circles (lower mobility than linear form) or linear molecules.

11.7.3 Difficulty in cloning deletion products

The success or otherwise of the exonuclease deletion can be monitored easily by agarose gel electrophoresis. Sometimes, however, apparently suitable deletions cannot be cloned. There are two likely explanations for this, other than the trivial ones of defective enzyme (polymerase, ligase) or other reagent.

Incomplete digestion with one of the two initial restriction enzymes will give substrate molecules for the exonuclease reaction which have either two resistant ends or two sensitive ends. Many restriction enzymes do not cut efficiently close to the end of a linear molecule so, if the two sites are close together, the second enzyme may work inefficiently, if at all. Information about the efficiency of digestion near the end of a DNA molecule is provided by some restriction enzyme suppliers in their catalog (e.g. New England Biolabs).

Molecules with two resistant ends will of course be unaffected by the exonuclease and will be religated to reform essentially the same molecule as the template. Conversely, molecules with two sensitive ends will be digested by the exonuclease from each end. This usually means that vector sequences will be deleted. If essential vector sequences (origin of replication, selectable marker) are destroyed by the endonuclease, these molecules cannot be recovered. In intermediate cases, where some substrate molecules have one sensitive and one resistant end, and others have two sensitive or two resistant ends, it is often possible to visualize the reaction products as distinct bands on an agarose gel – a resistant band at the original size, a band of deleted molecules and another band of higher mobility, representing molecules deleted from both ends. In such circumstances, it may be easier to purify the appropriate band from the gel, rather than attempting to optimize the exonuclease procedure.

The ends generated by the exonuclease/single strand nuclease procedure are rarely precisely blunt. Even a single protruding base will prevent re-circularization of the molecule. The final 'polishing' with Klenow or T4 DNA polymerase is therefore essential. However, this reaction does not appear to be very efficient in this context. In case of cloning difficulties, it may be helpful to purify the DNA away from contaminants (enzymes, DNA fragments, salts) by phenol/chloroform extraction and ethanol precipitation, then repeat the end-filling reaction.

11.7.4 Deletions using γδ transposon

Transposon insertion can be used to insert known sequence into an unknown region of DNA, as described in Section 11.6. It is also possible to use transposons to generate deletions [10]. Insertion of a transposon into a plasmid which already carries a copy of a transposon will generate a repeat. Direct repeats are converted to deletions by the host cell's recombination machinery. Many outcomes are possible other than the desired deletions, but a set of positive and negative selectable markers allows the selection of the desired clones (see *Figure 11.6*). The Deletion Factory™ kit from Life Technologies is based on this principle. This uses the γδ transposon to generate random deletions, then sorts them by use of built-in genetic markers and sizing on agarose gels. Deletions can be generated from either end of a template by using different combinations of markers. This system is intended primarily for generating deletion series in large (>5 kb) target DNAs, and has a *cos* site to allow the cloning of inserts of up to 40 kb.

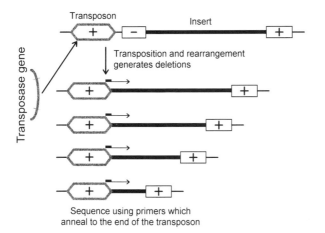

FIGURE 11.6: *Transposon-induced deletions. The insert (DNA of interest) is flanked by a transposon which has been engineered to carry a positive selection marker (+) a negative selection marker (–) and another, different, positive selection marker. Transposition by a cut-and-paste mechanism leads to a duplication of transposon sequences. This is resolved by recombination, leading to excision of one copy of the transposon and all the DNA in between. Selection against the negative marker and for both the positive markers allows the selection of those plasmids with a useful deletion. The size of the deletion is determined by PCR or restriction mapping, and selected clones are sequenced.*

References

1. Bankier, A.T., Weston, K.M. and Barrell, B.G. (1987) Random cloning and sequencing by the M13/dideoxynucleotide chain termination method. *Methods Enzymol.*, **155**, 51–93.
2. Sulston, J., Du, Z., Thomas, K., Wilson, R., *et al.* (1992) The *C. elegans* genome sequencing project: a beginning. *Nature*, **356**, 37–41.
3. Berg, C.M., Wang, G., Strausbaugh, L.D. and Berg, D.E. (1993) Transposon-facilitated sequencing of DNAs cloned in plasmids. *Methods Enzymol.*, **218**, 279–306.
4. Kasai, H., Isono, S., Kitakawa, M., Mineno, J., Akiyama, H., Kurnit, D.M., Berg, D.E. and Isono, K. (1992) Efficient large-scale sequencing of the *Escherichia coli* genome: implementation of a transposon- and PCR-based strategy for the analysis of ordered lambda phage clones. *Nucleic Acids Res.*, **20**, 6509–6515.
5. Strathmann, M., Hamilton, B.A., Mayeda, C.A., Simon, M.I. Meyerowitz, E.M. and Palazzolo, M.J. (1991) Transposon-facilitated DNA sequencing. *Proc. Natl Acad. Sci. USA*, **88**, 1247–1250.

6. Henikoff, S. (1987) Unidirectional digestion with exonuclease III in DNA sequence analysis. *Methods Enzymol.*, **155**, 156–165.
7. Alphey, L. (1995) DNA sequencing. in *DNA Cloning I – Core Techniques*, 2nd edn (D. Glover and B.D Hames, eds.) Oxford University Press, Oxford, pp. 225–260.
8. Sambrook, J., Fritsch, E.F. and Maniatis, T. (1989) *Molecular Cloning: A Laboratory Manual*, 2nd edn. Cold Spring Harbor Laboratory Press, Cold Spring Harbor, NY.
9. Wang, G., Blakesley, R.W., Berg, D.E. and Berg, C.M. (1993) pDUAL: a transposon-based cosmid cloning vector for generating nested deletions and DNA sequencing templates in vivo. *Proc. Natl Acad. Sci. USA*, **90**, 7874–7878.

12 Introduction to Bioinformatics and the Internet

Andy Brass, University of Manchester, UK

*"two months in the lab can easily save an afternoon
on the computer" Alan Bleasby*

12.1 Introduction

This section of the book is designed to demonstrate the ways in which bioinformatics techniques and resources can help both in managing sequencing projects and providing information on the structure and function of the gene/gene product being sequenced. The aim is not to give a thorough and extensive review of bioinformatics, but rather to give sufficient information to allow a molecular biologist to sensibly and *safely* use an appropriate range of analyses.

12.2 Bioinformatics is a knowledge-based theoretical discipline

Bioinformatics is a theoretical discipline – it attempts to make predictions about biological function using only sequence data. As such, it is a powerful tool in experimental design. However, it is important to realize that bioinformatics is fundamentally different from almost all other more 'mathematical' or 'axiomatic' theoretical disciplines.

Protein function cannot yet be *deduced* from an amino acid sequence, our knowledge of the fundamental mechanisms of molecular biology are still too simplistic to allow such deductions to be made. What we can do is ask whether any of the sequences whose function have been characterized resemble the unknown sequence in any way. The function of the unknown sequence can then be inferred from the type of similarities found. Most prediction in bioinformatics are made in a similar way, by comparing the unknown sequence against the biological knowledge base.

The key difference, therefore, between 'knowledge-based' and 'axiomatic' disciplines is the role played by the knowledge base of past experience. The challenge and the skill in bioinformatics is to make use of this knowledge base in the most effective way.

12.3 Access to bioinformatics tools

Recent revolutions in computer communication such as the Internet and the World Wide Web (Web), have made it much easier to access bioinformatics resources. For the price of a personal computer (PC) and an Internet connection, all the publicly available sequence databases and a wide range of analysis tools are instantly available free of charge to the bench molecular biologist. We are therefore assuming that you already have, or can easily get, access to the Internet and a Web browser (such as Mosaic, Netscape or Microsoft Internet Explorer). Many sites will also have access to bioinformatics software running on workstations using the UNIX operating systems. The Genetics Computing Group (GCG) provide an extensive suite of bioinformatics programs to run on such workstations. However, although mention will sometimes be made of such packages where there is no Web-based alternative, the main thrust of the book will be to concentrate on the most widely available and accessible set of tools, that is those on the Web.

12.3.1 Getting access to tools on the Web

Given the ubiquitous nature of the Web, we are assuming that people are used to using a standard Web browser (typically either Netscape or Microsoft Internet Explorer). For those not used to using these facilities, there are a large number of straightforward introductory books available.

However, to make full use of the bioinformatics facilities available on the Internet, there are a number of other Internet facilities that users should be aware of and these will be explained below.

12.3.2 Navigating the Web – or how do I find what I want?

The Web is enormous and growing rapidly – it is often a non-trivial task to find the resources needed for a specific task. However, there are a number of strategies which can be used to help you find what you want.

Sites with lists of bioinformatics resources. Firstly, many people have produced Web pages which attempt to collate bioinformatics resources. By going to one of these pages, you should rapidly be able to find the most commonly accessed bioinformatics resources. A number of such pages and their addresses are listed in *Table 12.1* – these pages are well worth adding to your list of favorite bookmarks. The links given below are only a small number of those available – any search of the Net (see below) will find many others which you may find more useful.

Web search engines. New bioinformatics Web resources are being produced all the time, and so the lists of bioinformatics resources

TABLE 12.1: *Links to Web pages containing lists of bioinformatics tools and resources on the Web*

Internet address	Comments
http://www.public.iastate.edu/~pedro/research_tools.html	Pedro's Biomolecular research tools – a well-known resource but getting rather out of date
http://expasy.hcuge.ch/www/amos_www_links.html	Amos Bairoch's personal collection of favorite sites – including probably the most comprehensive collection of links to databases and database resources
http://expasy.hcuge.ch/tools.html	A listing of sequence analysis tools
http://ahsc.arizona.edu/~lei/genetics/tools.htm	A well-structured resource with extensive links to other sites, descriptions of resources and Web search facilities
http://www.incyte.com/Globe/bioinfo.html	Well-structured and comprehensive
http://www.ebi.ac.uk/htbin/bwurld.pl	A search engine to find biologically relevant Web sites

might not yet reference a useful resource. There are a large number of different search engines available on the Internet that allow you to hunt for resources on the Web by keywords. Some of the more useful such engines and their addresses are listed in *Table 12.2*. Be aware that none of these engines is particularly sophisticated, and you will often need to wade through large numbers of irrelevant (but often entertaining) hits to find the something of use. Infoseek is useful in that it grades hits on a percentage match basis.

TABLE 12.2: *Internet addresses of Internet search tools*

Address	Comments
http://altavista.digital.com/	A useful site in that it also allows you to search newsgroups
http://lycos.cs.cmu.edu/	Another good tool for searching for keywords
http://www.yahoo.com/	Splits sites into related areas
http://www.infoseek.com	Grades hits on a percentage match basis
http://www.excite.com	Concept-based searching
http://www.webcrawler.com	Supports natural language searching

If one of these tools does not find exactly what you want, it is worth trying some of the others – different tools tend to access slightly different sets of data.

Newsgroups. Newsgroups are an important way of keeping up to date with developments in bioinformatics. To access newsgroups, you need special news reader software and access to a news server, that is a site from which you can access the newsgroups. News reader software often comes bundled as part of an Internet browser, for example Netscape and Microsoft's Internet Explorer both contain news readers. News reader software is also available from a number of sites on the web, for example http://www.forteinc.com. If you receive your Internet access through a service provider then you should have been given access to a news server along with your access package. If you are trying to access the newsgroups from a University site, then the local computing center would normally run a news server to which you can attach.

A number of newsgroups found under the *bionet.* grouping are available which are excellent sources of bioinformatics information. A selection of these groups is listed in *Table 12.3*. If you are having serious problems either finding or using a particular resource, then questions asked on a group such as *bionet.software* normally get you a quick and knowledgeable response. If a newsgroup is moderated it means that all postings to the group are checked by the moderator before being posted. This has the effect of significantly reducing the amount of irrelevant discussion, junk mail sent to that group. To see the effect of this, compare the postings in bionet.general (unmoderated), with those in bionet.announce (moderated).

TABLE 12.3: *Newsgroups containing information relevant to bioinformatics*

Newsgroup	Contents
bionet.announce	Announcements of widespread interest to biologists (moderated)
bionet.biophysics	The science and profession of biophysics
bionet.celegans	Research discussion of the organism *Caenorhabditis elegans*
bionet.general	General biosciences discussion
bionet.glycosci	Research issues re carbohydrate and glycoconjugate molecules
bionet.immunology	Discussions about research in immunology
bionet.infotheory	Discussions about biological information theory
bionet.jobs	Scientific job opportunities
bionet.software	Information about software for biology

There is a problem with newsgroups in that messages and their replies are only stored for a relatively short length of time on the news server before being deleted. However, all messages posted to the bionet newsgroups are archived and stored at http://www.bio.net. By choosing the 'Access the BIOSCI/bionet Newsgroups' option, you can then search a particular newsgroup to see whether a particular topic has been discussed before. This is probably one of the most useful of all the information resources available for bioinformatics on the Web. You will not be the first person to encounter most problems – more often than not someone else has had the same problem, asked the question in one of the bionet newsgroups and received a useful answer. If the answer you want is not in the archives, then you can safely submit the question to the newsgroup knowing that it has not been asked 20 times before.

12.3.3 Using Web-based tools

Using most Web-based tools is straightforward and simply requires a user to be able to use a mouse and understand how to cut and paste text. For example, if you want to use a Web tool to search for similarities to your favorite sequence in a database you would:

- load your sequence into a word processor package such as Microsoft Word;
- start up your Internet browser and go to the address of the appropriate resource – you should see a box into which to paste the sequence;
- highlight the sequence in the word processor by holding down the shift key and dragging the mouse over the sequence. Press [control]-c to copy the sequence into a buffer;
- move the mouse across to the text box on the Web page and type [control]-v to paste the sequence into the box.

A similar technique can be used to paste results from a Web search back into a word processor, or alternatively use the Save option in the browser's file menu.

12.3.4 E-mail servers

Several types of bioinformatics analysis cannot be run quickly, for example comparing a sequence against a database. It simply is not practical to try to run this type of job using a real-time Web interface. Firstly, users would be waiting at their browser screen too long for results to come back, and secondly it could be placing considerable demands on the machine running the analysis. It makes more sense to run such resource-hungry applications at times when the host machine has the appropriate spare capacity. This can be achieved using e-mail servers. Essentially, the request to run the job is sent as an e-mail message to the machine running the service. This request can then automatically be placed in a queue to run when resources become available. The results of the analysis can then be returned as a mail message to the user. A number of key bioinformatics tools can be accessed in this way, including database searches and structure prediction programs.

It is important that the mail containing the job request is properly formatted because it will be handled automatically. Almost all e-mail servers (see *Table 12.4*) will send you a manual describing how they should be used if you first send the e-mail server an e-mail containing simply the word 'help' as the message text.

12.3.5 Accessing remote computers to get useful software – anonymous ftp

You will often see reference to bioinformatics software available over the Web, particularly in newsgroups such as bionet.software. For example, there are now a number of tools that allow you to edit multiple sequence alignments on a PC with a nice, intuitive windows interface. Normally, the person who has developed the software will have made an announcement somewhere like the newsgroup bionet.software explaining what the software does and giving the address of an ftp site from where the software can be obtained (see *Table 12.5*) (ftp stands for file transfer protocol, and is the main method used for transferring files across the Internet – you should have software for using ftp included along with the general software allowing you access to the Internet).

Protocol 12.1 describes how to get a working version of the program rasmol from the ftp site ftp.dcs.ed.ac.uk. Rasmol, written by Roger Sayle from Glaxo Wellcome, is a nice utility which allows you to view

TABLE 12.4: E-mail address of mail servers offering useful bioinformatics tools

E-mail address	Short description
blast@ncbi.nlm.nih.gov	Submit BLAST database search
blitz@ebi.ac.uk	Submit database searches using the Smith–Waterman algorithm
geneid@bir.cedb.uwf.edu	Gene identification from DNA sequence data
grail@ornl.gov	Prediction of exons from genomic DNA
mowse@dl.ac.uk	Identification of proteins from mass spectroscopic analysis of protease fragments
nnpredict@celeste.ucsf.edu	Neural network secondary structure prediction
predictprotein@embl-heidelberg.de	Protein secondary structure prediction – probably the most accurate techniques available
pythia@anl.gov	Allows the identification of human repetitive DNA elements such as L1, MERx, LTR, etc. Can also be used to identify the presence of Alu sequences and to classify them into subfamilies
signalp@cbs.dtu.dk	Prediction of signal peptides in protein sequences
tmap@embl-heidelberg.de	Prediction of transmembrane helices from protein sequence data

A more comprehensive listing of available services can be found on the expasy server in the file ftp://expasy.hcuge.ch/databases/info/serv_ema.txt

TABLE 12.5: Anonymous ftp sites containing useful bioinformatics software or databases

Address	Comments
ftp://ftp.ebi.ac.uk/pub/software/	A large collection of software and databases from the European Bioinformatics Institute
http://ncbi.nlm.nih.gov/Home.html	A Web address to the ftp site for NCBI
http://iubio.bio.indiana.edu/	An extensive collection of software for UNIX, Mac and DOS

A more comprehensive listing of sites is available in the file ftp://expasy.hcuge.ch/databases/info/serv_ftp.txt

and rotate three-dimensional images of protein structures on a PC or Mac – it also allows you to calculate and display protein backbones, secondary structures, etc. It is therefore an indispensable tool for anyone interested in protein structure, and is available free over the Internet. The ftp site for rasmol was found by using Alta Vista (see

Section 12.3.2) with rasmol as the search word – it also provided the information that the rasmol software is in the directory pub/rasmol.

12.4 Good and bad practice

Most bioinformatics tools on the Web use computers and computing resources that other people have paid for – services are being provided as a favor to the community, not as a right. This places two main obligations on you as a user. Firstly, do not abuse the system by swamping it with large numbers of jobs all at the same time. Secondly, if you have developed useful software, or have created useful data sets, think about creating your own Web site to make them available to the community – give as well as take.

Many naïve users have difficulties with newsgroups, either posting inappropriate questions to the group or asking questions that have been asked many times before. Newsgroups often produce regular faqs (a message of frequently asked questions) – unless you know a group well it helps to check this file to see if the answer to your question is contained within, alternatively check the bionet archives on www.bio.net. Do what you can to keep the signal to noise ratio on the bionet newsgroups as high as possible.

The most productive time to use the Web is when the US is asleep. In Europe this means the best time to use the Web is in the morning – by the afternoon the traffic on the Web becomes noticeably higher and performance slows significantly. US-based users just have to get up very early.

PROTOCOL 12.1: Obtaining software from an anonymous ftp site

1. You can connect to the ftp site using Netscape or other browsers by connecting to the address ftp://ftp.dcs.ed.ac.uk. Using ftp in the URL as opposed to the more traditional http allows you to access the anonymous ftp server without needing to type in a login name and password. Next double click on the appropriate folders until you reach the required zip files. Clicking on these files will bring them back to your machine. Netscape will prompt you for the file name under which you want to store the files. Finally uncompress them as above.

2. Uncompress the files.

   The files raswin.zip and raswin.zip should now be on your hard disk. Create a new directory and place these two files into it. To uncompress them use the pkunzip program (available from almost all ftp sites if you do not already have a copy)

   c:\>pkunzip -d rasmol.zip
   c:\>pkunzip -d rasmenu.zip

   If you now go back to windows and examine the directory containing the zip files should see the rasmol icon. Now simply double click and have a play with the software.

---

13 Sequence Databases

Andy Brass, University of Manchester, UK

13.1 Background

Bioinformatics databases can be split broadly into two types, primary and secondary. Primary databases contain original biological data such as DNA sequence, or protein structure information from crystallography. Secondary databases attempt to add value to the primary databases and make them more useful for certain specialist applications, for example the EPD database of eukaryotic promoter sequences, or PROSITE, the database of common structural or functional motifs found in proteins.

A database entry typically consists of two parts: the raw sequence data, and an annotation describing the biological context of the data. It is important to realize that the information contained in the annotation is just as important and useful as that in actual sequence data. The advent of genome scale sequencing is also creating problems in annotation. Often very little is known about the sequences coming off the automatic sequencing machines other than the cell type from which they came. If you are trying to determine the function of an unknown protein sequence, it can be very frustrating if you find a good match only to discover that there is no information describing the function of this protein.

Different databases contain differing quality of annotation information and there is often a trade off between the completeness of the data and the work put into annotation – some databases provide excellent coverage of the sequence data but at the expense of the annotation, other databases cover less of the data but provide much more complete annotation. Formats and conventions for the type and completeness of the data entered in the annotation are still in a state of flux, therefore much of the information on annotation in this section may soon become outdated. It is also worth bearing in mind that all bioinformatics databases contain a small percentage of entries in which either the data or the annotation is wrong.

13.2 Primary databases

13.2.1 DNA databases

The major primary data resource is DNA sequence. Three groups now collaborate worldwide to produce computerized DNA sequence databases (see *Table 13.1*): the European Molecular Biology Laboratory (EMBL) (based at Cambridge, UK), GenBank [based at the National Center for Biotechnology Information (NCBI), a division of the National Library of Medicine, located on the National Institutes for Health (NIH) campus in the US], and the DNA Databank of Japan (DDBJ). These groups exchange information daily and have agreed common standards for DNA sequence database entries. In essence, each organization is responsible for collecting data from a different geographical area (EMBL covers Europe, Genbank covers the Americas, etc.) and then all the information from these different locations is pooled and shared between the three databases.

TABLE 13.1: Web addresses to sites containing nucleotide data

Database	Address
EMBL	www.ebi.ac.uk/ebi_docs/embl_db/ebi/topembl.html
GenBank	www.ncbi.nlm.nih.gov/Web/Search/index.html
DDBJ	www.ddbj.nig.ac.jp

The growth of the DNA sequence databases has been phenomenal. Release 49 of EMBL (released in November 1996) contained over 696 183 709 nucleotide bases from 1 047 263 different sequences. Human entries predominate, constituting 54% of the total. However more than 15 500 species are represented with the next five most popular species (in terms of bases) being *Caenorhabditis elegans*, *Saccharomyces cerevisiae*, *Mus musculus* and *Arabidopsis thaliana*. Historically, the database has been doubling in size every 22 months, but that rate has accelerated rapidly due to the enormous growth in data from ESTs. The current doubling time for EMBL is now down to around 9 months. This speed of growth has important consequences for database search strategies. It could well be that if you do not find a match to your sequence in one month, you could find one in the next database release. Similarly, when writing up the results of a Bioinformatics analysis it is very important to state which releases of the sequence databases were used.

Because of the size of the databases, they are now broken up into divisions defined primarily by species. For example, the EMBL database currently contains separate sections for ESTS, bacteriophages, invertebrates, organelles, primates, rodents, other vertebrates, other mammals, plants, prokaryotes, etc. These divisions are useful in that they make it easier to search only in the relevant part of the database. The definition of these divisions has not yet settled and could well be different by the time you read this – primarily because of the pressures caused by the rapid increase in size of the databases. The latest division to be added is one for the results of high-throughput screening experiments. The user manuals for EMBL and GenBank can be found at http://www.ebi.ac.uk/ebi_docs/embl_db/usrman/usrman.html and ftp://ncbi.nlm.nih.gov/genbank/gbrel.txt respectively, and these provide more detail on the exact way the databases are organized.

13.2.2 Genome databases

A second major source of primary data is the various genome projects. A number of genome projects have been completed, included a representative eukaryote (*Saccharomyces cerevisiae*), an archeon (*Methanococcus janeschii*) and three prokaryotes (*Haemophilus influenzae*, *Mycoplasma genitalium* and *Escherichia coli*). Much of the information from these projects can now be found in EMBL. A large number of genome projects are now underway. The information contained within these sources is outside the scope of this section, but *Table 13.2* shows the Web addresses of a representative sample of these resources.

13.2.3 Protein sequence databases

Two main protein sequence databases have grown up to mirror the EMBL and GenBank databases. The SWISS-PROT database consists of properly checked and annotated translations of sequences in the EMBL database. It is maintained collaboratively by the Department of Medical Biochemistry at the University of Geneva and the European Bioinformatics Institute (EBI), predominantly. *Figure 13.1* shows the growth in the number of residues stored within the SWISS-PROT database as a function of time. Unfortunately there are delays in translating EMBL entries into properly annotated protein translations. A significant number of DNA entries with published

TABLE 13.2: Web addresses for a sample of sites containing information from the genome projects

Organism	Address
Mouse	http://www.informatics.jax.org/mgd.html
Rat	http://ratmap.gen.gu.se
Dog	http://mendel.berkeley.edu/dog.html
Cow	http://locus.jouy.inra.fr/cgi-bin/bovmap/intro2.pl
Pig	http://www.ri.bbsrc.ac.uk/pigmap/pigbase/pigbase.html
Sheep	http://dirk.invermay.cri.nz
Chicken	http://www.ri.bbsrc.ac.uk/chickmap/chickgbase/manager.html
Zebra fish	http://zfish.uoregon.edu/
Caenorhabditis elegans	http://www.ddbj.nig.ac.jp/htmls/celegans/html/CE_INDEX.html
Dictyostelium discoideum	http://glamdring.ucsd.edu/others/dsmith/dictydb.html
Drosophila	http://morgan.harvard.edu
Mosquito	http://klab.agsci.colostate.edu
Arabidopsis	http://genome-www.stanford.edu/Arabidopsis
Cotton	http://algodon.tamu.edu
Beans	http://scaffold.biologie.uni-kl.de/Beanref
Maize	http://www.agron.missouri.edu
Rice	http://www.staff.or.jp
Soya	http://mendel.agron.iastate.edu:8000/main.html
Trees	http://s27w007.pswfs.gov

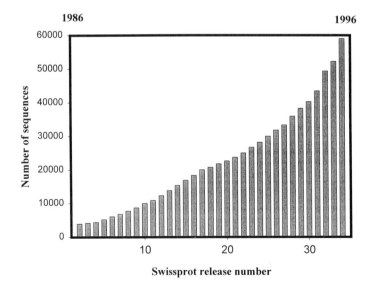

FIGURE 13.1: The growth in the number of sequences in the SWISS-PROT databases over the past decade.

open reading frames (ORFs) are not yet in SWISS-PROT. To help overcome this problem, TREMBL (Translated EMBL) has recently become available. TREMBL is a protein database which contains all the protein-coding regions described in EMBL. This provides a very comprehensive coverage of protein sequences but at the expense of the quality of the annotation.

The PIR databases are created by the NCBI as translations of GenBank. The PIR database is split into four sections in terms of the degree of annotation provided, as shown in *Table 13.3*.

TABLE 13.3: *The divisions in the PIR database (release 51.03)*

Name	Comment	Number of entries
PIR1	Classified and annotated entries	13 572
PIR2	Annotated entries	69 368
PIR3	Unverified entries	7 508
PIR4	Unencoded or untranslated entries	196

Web addresses and contact points for the protein sequence databases are shown in *Table 13.4*. Full details of SWISS-PROT and the officially sanctioned nomenclature can be found at http://expasy.hcuge.ch/txt/userman.txt and http://expasy.hcuge.ch/sprot/sp-docu.html

TABLE 13.4: *Web addresses of sites containing protein sequence data*

Database	Address
SWISS-PROT	http://expasy.hcuge.ch/sprot/sprot-top.html
TREMBL	http://www.ebi.ac.uk/pub/databases/trembl
PIR	http://www-nbrf.georgetown.edu/pir/

13.2.4 Protein structure databases

Experimentally derived three-dimensional structures of proteins are stored at the protein databank (PDB) (http://pdb.pdb.bnl.gov/). Although there are not as many entries in PDB as in the protein sequence databases, the number is increasing exponentially. Structures determined by both X-ray and nuclear magnetic resonance (NMR) are stored. The NRL-3D database provides a database of the sequences of proteins in PDB. This database allows sequence comparisons to be made just against the subset of proteins of known structure (see Section 17.2 for more details).

13.3 Primary sequence database annotation

At present, there is no common format for sequence annotation between the different database centers. Although, in principle, the same information is provided, the format in which it presented varies. A more serious concern is that different annotation information is provided for DNA and protein sequence database entries for the same gene. An example of the annotation for an entry in the EMBL database is shown in *Figure 13.2* and the datafields are explained in *Table 13.5*.

The database cross-reference field needs a more detailed explanation. Many secondary databases access information from the primary

```
ID   LISOD        standard; DNA; PRO; 756 BP.
XX
AC   X64011; S78972;
XX
DT   28-APR-1992 (Rel. 31, Created)
DT   30-JUN-1993 (Rel. 36, Last updated, Version 6)
XX
DE   L.ivanovii sod gene for superoxide dismutase
XX
KW   sod gene; superoxide dismutase.
XX
OS   Listeria ivanovii
OC   Prokaryota; Bacteria; Firmicutes; Regular asporogenous rods.
XX
RN   [1]
RA   Haas A., Goebel W.;
RT   "Cloning of a superoxide dismutase gene from Listeria ivanovii by
RT   functional complementation in Escherichia coli and
RT   characterization of the gene product.";
RL   Mol. Gen. Genet. 231:313-322(1992).
XX
RN   [2]
RP   1-756
RA   Kreft J.;
RT   ;
RL   Submitted (21-APR-1992) on tape to the EMBL Data Library by:
RL   J. Kreft, Institut f. Mikrobiologie, Universitaet Wuerzburg,
RL   Biozentrum Am Hubland, 8700 Wuerzburg, FRG
XX
DR   SWISS-PROT; P28763; SODM_LISIV.
XX
FH   Key             Location/Qualifiers
FH
FT   source          1..756
FT                   /organism="Listeria ivanovii"
FT                   /strain="ATCC 19119"
FT   RBS             95..100
FT                   /gene="sod"
FT   CDS             109..717
FT                   /gene="sod"
FT                   /EC_number="1.15.1.1"
FT                   /product="superoxide dismutase"
FT                   /db_xref="PID:g44011"
FT                   /db_xref="SWISS-PROT:P28763"
FT                   /translation="MTYELPKLPYTYDALEPNFDKETMEIHYTKHHNIYVTKLNEAVSG
FT                   HAELASKPGEELVANLDSVPEEIRGAVRNHGGGHANHTLFWSSLSPNGGGAPTGNLKAA
FT                   IESEFGTFDEFKEKFNAAAAARFGSGWAWLVVNNGKLEIVSTANQDSPLSEGKTPVLGL
FT                   DVWEHAYYLKFQNRRPEYIDTFWNVINWDERNKRFDAAK*"
FT   terminator      723..746
FT                   /gene="sod"
XX
SQ   Sequence 756 BP; 247 A; 136 C; 151 G; 222 T; 0 other;
```

FIGURE 13.2: *The annotation for an EMBL entry (accession code X64011). A brief explanation of the terms in the annotation is given in* Table 13.5.

TABLE 13.5: An explanation of the terms appearing in the EMBL annotation

Code	Full meaning	Comments
ID	identifier	The first entry on this line is the name of the database entry. Other information refers to the status of the entry (whether it has been properly checked – standard in this case), and the number of bases in the entry
AC	accession number	This is the unique identifier for the entry. It will never be changed. If two entries are merged into one then both original accession numbers will be shown
DT	date	Two dates shown, one for the first time the data appeared, one for the latest revision
DE	description	A free text description of the gene
KW	keywords	Words describing the gene product
OS	organism (species)	The name of the organism
OC	organism (classification)	A brief taxonomy for the species. This taxonomy is not truly rigorous and should be treated with caution
OG	organelle	Is the gene found in a specific organelle?
RN	reference number	A series of fields used to store reference information relating to the database entry
RC	reference comment	
RP	reference positions	
RX	cross-reference	
RA	reference authors	
RT	reference title	
RL	reference location	
DR	database cross-reference	See text
FH	feature table header	A header for the feature table
FT	feature table data	See text
CC	comments	Free text comments on the entry
XX	spacer line	
SQ	sequence header	Information on the size and composition of the sequence
blank	sequence data	At last – the sequence data itself
//	termination line	Marks the end of an entry

databases. For example the database OMIM (Online Mendelian Inheritance in Man) contains entries relating to human genetic disease. If an entry in OMIM refers to a known sequenced gene, that gene will be referred to by its accession number. If this were the case, the DR line would contain the keyword OMIM and the name of the OMIM entry which referred to the EMBL sequence. In *Table 13.5* the only entry in DR is to the protein translation of this gene in SWISS-PROT. Also given on this line are the accession number and identifier of the SWISS-PROT entry. The DR line is therefore a very powerful tool for helping to make links from the original database to other databases containing information on the sequence in question. *Table 13.6* lists some of the databases likely to be encountered in the DR line.

The other important field in the annotation which needs further explanation is FT, the features table. The features table is an attempt to get across as much information as possible on the sequence and what it does in a computer-readable form. The three main DNA databases (EMBL, GenBank and DDBJ) have agreed a common language with which to describe these features. This language is described in full at http://www.ebi.ac.uk/ebi_docs/embl_db/ft/feature_table.html

A number of DNA entries contain more than one ORF. The PID identifier in the features table is used to identify each ORF uniquely. This is an important new addition to the annotation information as it makes it possible, for example, to link a number of different SWISS-PROT entries to the same EMBL sequence and know precisely which ORF in the EMBL sequence corresponds to which SWISS-PROT entry.

TABLE 13.6: *Some of the databases cross-referenced in the sequence databases via the DR line*

Database	Comment
SWISS-PROT	Protein sequence database
EMBL	DNA sequence database. In a protein database the DR line would point to the DNA entry from which it had been translated
OMIM	Mendelian Inheritance in Man – provides information on whether the sequence has been implicated in a genetic disease
PROSITE	A database of conserved protein motifs (see below)
HSSP	A database of proteins homologous to proteins of known structure – useful for determining the fold of your protein
PDB	Protein structure database – does the sequence have a known structure
MEDLINE	The abstract of the published paper referred to in the RL line
PIR	The equivalent protein entry in PIR

13.4 Information retrieval systems

There are a number of systems which provide users with simple, intuitive access to the data in sequence databases. The two best known and most accessible are Entrez (developed in the US) and SRS (Sequence Retrieval System), developed by Theore Etzold at EMBL.

The SRS software is used at a number of sites across Europe. SRS is a flexible system and can be used to access a number of different databases. This means that the databases accessible using SRS will vary slightly from site to site, depending on which databases have been made available by the local system administrator. For example, the OWL database is a nonredundant database of protein sequences created by collecting together all the nonredundant protein sequence entries from all the major protein databases. The OWL database is accessible and can be searched using the SRS service at SEQNET (http://www.seqnet.dl.ac.uk/srs/srsc). However, OWL is not available from SRS at the EBI (see below).

Sequences are normally retrieved either via an accession number (e.g. from a published paper) or through information contained within the sequence annotation. The advantage of using SRS is that it allows all the major databases to be accessed through a common front end and with the appropriate hot links in place to entries in the DR (database cross-reference) fields. SRS therefore allows you to search a wide variety of databases using a common interface and to follow the links to related information in other databases.

Using SRS is straightforward. *Figure 13.3* shows the first screen shown when connecting to SRS at the EBI (http://www.ebi.ac.uk/srs/srsc). To retrieve sequence data you would choose the first option, 'Search sequence libraries', by clicking the mouse on the button. This presents a new screen (*Figure 13.4*). Click on the database buttons to choose the databases you wish to search and then enter appropriate search terms in the text boxes below. Note that it is possible to connect search terms using a variety of logical operators (and, or, not) and to constrain a given search term to a particular field of the annotation information. The search is started by clicking on the 'DO-QUERY' button. Additional options are available to modify either the format in which the data is returned or the information given on the entries found. Useful entries can be saved to your own computer using the save options on your Internet browser. More information on SRS can be obtained from the manual (available at many sites, including http://www.sanger.ac.uk/srs/srsman.html).

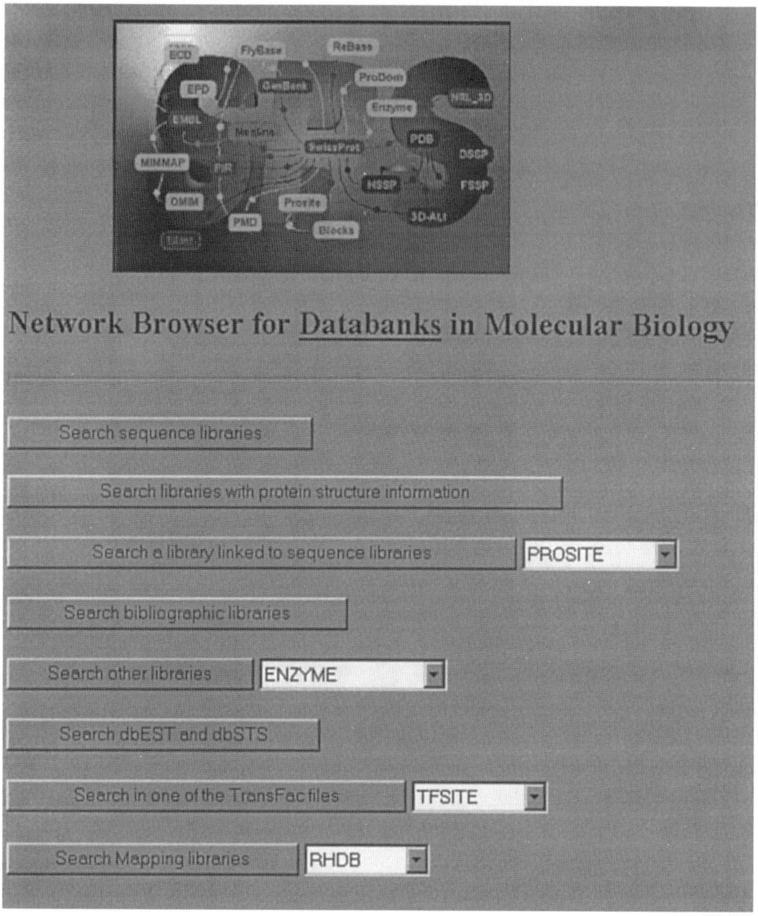

Network Browser for <u>Databanks</u> in Molecular Biology

Search sequence libraries

Search libraries with protein structure information

Search a library linked to sequence libraries | PROSITE ▼

Search bibliographic libraries

Search other libraries | ENZYME ▼

Search dbEST and dbSTS

Search in one of the TransFac files | TFSITE ▼

Search Mapping libraries | RHDB ▼

FIGURE 13.3: The initial page of SRS. To search the DNA or protein sequence databases for an entry you would click on the 'Search sequence libraries' button.

13.5 Submitting a sequence to a database

Many journals will not accept a paper containing sequence data unless the sequence has first been deposited in one of the sequence databases. The way in which sequences should be deposited depends on where you are based; sequences generated in Europe being deposited with EMBL, sequences generated in the Americas deposited in GenBank, and those generated in Japan or the Pacific Rim being sent to DDBJ. The home page for each of these databases will contain

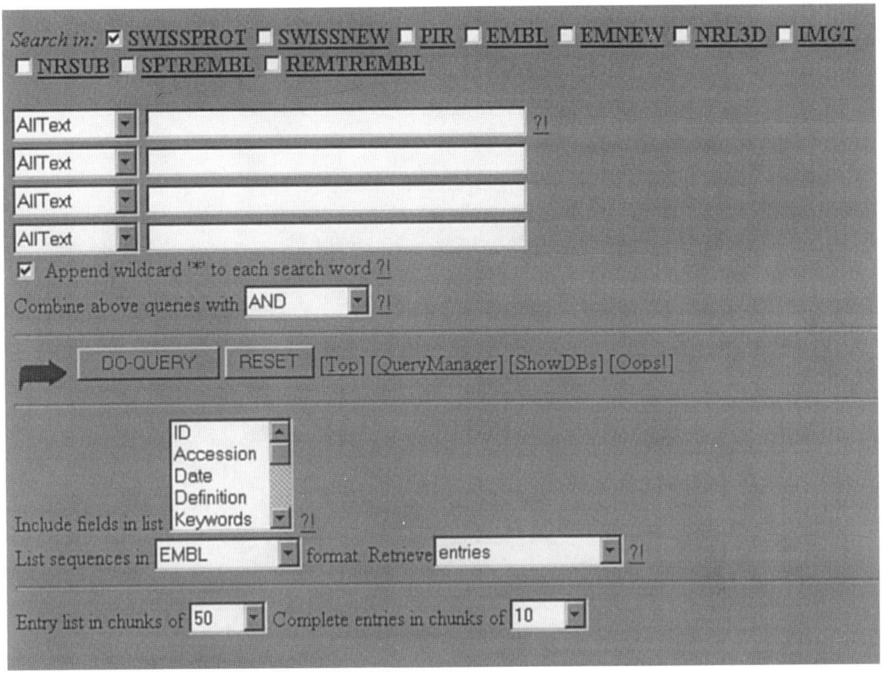

FIGURE 13.4: The SRS page for searching the sequence database. To search a specific database you click on the database you require (in this case SWISS-PROT) and then enter the appropriate search terms in the text boxes. To submit the query you must click on the 'DO-QUERY' button.

details on how to deposit data electronically. For example, a Web form to submit entries to EMBL can be found at http://www.ebi.ac.uk/subs/emblsubs.html. Once received, the data will be given an accession number which can be referenced in publication. There will be a delay before the sequence appears in the full database to allow the database staff to produce a fully annotated entry and perform a degree of data checking on the entry.

14 Sequence Alignment and Databases Searches

Andy Brass, University of Manchester, UK

14.1 Introduction

Sequence alignment attempts to align two or more sequences (DNA or protein) such that regions of structural or functional similarity between the molecules are highlighted. By asking whether our unknown sequence is in any way similar to known sequences (and ideally to sequences of known function or structure), we can identify the unknown sequence and predict its structure and function.

Software is available on the Web to perform a wide range of different types of sequence alignment. However, if the results returned from these programs are to be useful, it is important that they are not just used as black box programs. Sequence alignment is a nontrivial computational task and the choice of program and program parameters can have a significant impact on the sensitivity of the final results. Missing a weak, but significant match because of an inappropriate parameter choice can be a very expensive mistake to make – not least in experimental time.

14.2 Scoring matrices

Calculating the optimal alignment between two sequences is not straightforward. We need to find a way of calculating a number for any alignment such that largest score corresponds to the most biologically significant match.

145

Consider the two alignments of amino acids shown below (see *Table B.2* for one-letter code). If we just scored alignments by the number of residues in common then both would be scored equally (with five matches out of nine):

(a)   TTYGAPPWCS   (b)   TTYGAPPWCS
       TGYAPPPWS   TGYAPPPWS
       *   * * *   *   * *   * * *

However, whereas alignment (a) only conserves relatively common residues (A, P, S and T), alignment (b) conserves less common residues such as W (tryptophan) and Y (tyrosine). We need a method for scoring matches between amino acids which reflects their biological and chemical relationships.

A C-to-C match in alignment is more important than an S-to-S match because cysteine is a relatively rare amino acid with very particular properties whereas serine is relatively common. Similarly a D-to-E match should score positively because the two residues are chemically similar and could be doing the same job in the two proteins being aligned. However, a V-to-K match should be penalized because the two residues are so dissimilar it is unlikely that they could be doing the same job in the two proteins.

The scoring matrix contains the information on how to score each of the matches in the alignment. Scoring matrices for DNA alignments are relatively straightforward. A commonly used one is shown below:

	A	**C**	**G**	**T**
A	0.9	−0.1	−0.1	−0.1
C	−0.1	0.9	−0.1	−0.1
G	−0.1	−0.1	0.9	−0.1
T	−0.1	−0.1	−0.1	0.9

that is every base pair match in the alignment scores 0.9, every mismatch is penalized by a score of −0.1. So, for example, the alignment shown below would receive a score of 4.3 [= (5×0.9) + (2×−0.1)].

                GCGCCTC
                GCGGGTC
                * * *   * *

The situation for scoring matrices for use with proteins is slightly more complicated as there is no one matrix which can be used universally. The scoring matrix attempts to reflect the ways in which one amino acid can mutate into another over the evolutionary history of a particular protein family. Different scoring matrices are appropriate depending on the protein family involved and the degree of similarity expected.

The two best known classes of scoring matrix likely to be encountered are the PAM matrices developed in the 1980s by Margaret Dayhoff [1] and the more recent BLOSUM set of matrices developed by Henikoff and Henikoff [2] (see *Table 14.1*). The PAM matrices were built from an alignment of cytochrome proteins. The BLOSUM matrices have been built upon the more extensive set of alignments that can now be generated with the increase in size of the sequence databases.

TABLE 14.1: *The standard scoring matrices*

Scoring matrix	Comment
BLOSUM62	Good, robust general purpose matrix – the default choice
PAM40	Good for detecting sequences which have only diverged recently
PAM120	Another general purpose matrix
PAM250	Used for detecting sequences which diverged a long time ago (~ 25% sequence identity) – has a reputation for not being very selective

Finally, an important point of semantics in sequence alignment. Many people use the terms homology and similarity interchangeably – this is a mistake. To describe two sequences as being homologous implies that there is an evolutionary relationship between them, they evolved from a common ancestor sequence, to describe them as similar just implies that their sequences are similar – the two terms are not interchangeable.

14.3 Gap penalties

A second important set of parameters in sequence alignment controls the costs that are given to insertions and deletions. As sequences evolve they can collect insertions and deletions. Sometimes just one or two resides are involved in the insertions or deletions, other times entire domains can be inserted. The scoring system used to penalize gaps in the sequence must reflect this.

There are two parameters used for gap penalties, one for gap opening the other for gap extension. The gap opening penalty penalizes the creation of a single gap, the gap extension parameter is the cost that must be paid for increasing the size of a gap once opened. Mathematically the penalty, W_k, paid for opening a gap of length k is written as:

$$W_k = a + bk$$

where a is the gap opening penalty and b is the gap extension penalty. The effect of varying these values is outlined in *Table 14.2*. Unless you know in advance the sort of match you are looking for, it is useful to run alignments with a couple of different values for the gap penalty parameters, one set to allow for a few large insertions, another to allow for more small insertions and deletions.

TABLE 14.2: *The effect of varying gap creation and extension parameters on the type of alignment produced*

Gap opening penalty	Gap extension penalty	Comment
Large	Large	Very few insertions or deletions – useful to obtain good alignments for very closely related proteins
Large	Small	A few large insertions – useful for situations where an entire domain might have been inserted
Small	Large	Many small insertions – useful for more distantly related homologous proteins

14.4 Pairwise sequence alignments

There are two well-known algorithms for calculating the optimal alignment between two sequences. The Needleman–Wunsch algorithm [3] calculates a global alignment between two sequences – the best alignment including all of the shortest sequence. The Smith–Waterman algorithm [4] is a modification of Needlemann–Wunsch that returns the best local alignment between two sequences. Both algorithms can be used on either protein or nucleic acid sequences. Both algorithms will always return the alignment with the maximum possible alignment score for a given choice of gap penalties and scoring matrix. However, the alignment returned need not be biologically significant (see Section 14.7). A large amount of software is available via anonymous ftp for pairwise sequence alignment. Within GCG, the programs BESFIT and GAP calculate pairwise alignments. On the Web there are only a limited number of sites that allow you to align two sequences:

Resource name: ALIGN
Address: http://genome.eerie.fr/fasta/align-query.html
Comments: attempts to calculate an optimal alignment between two sequences provided by a user. Allows a choice of scoring matrices to be used, but does not allow gap penalties to be varied.

Resource name: Align
Address: http://www.mips.biochem.mpg.de/mips/programs/align.html
Comments: only allows alignments of sequences already on the databases. You cannot submit your own sequences to this service.

14.5 Multiple sequence alignments

It is possible to prove that the pairwise alignment algorithms mentioned above always return the best possible match between two sequences. Once we try to align more than two sequences at once this is no longer the case. Any automated multiple sequence alignment can only give a first approximation to an alignment – you should always check the alignment yourself to ensure that it cannot be improved. The human eye (coupled to biological insight) is much better at spotting patterns in multiple sequence alignments than any computer can so far achieve. A number of tools are available for multiple sequence alignment on the Web, for example the recent CINEMA program [5], which not only aligns the sequences but allows for easy editing of the program output using a friendly point and click interface. CINEMA illustrates the power of the Java programing languages in a bioinformatics environment.

Multiple sequence alignments of homologous sequences are a powerful tool in the bioinformatics armory. They can be thought of as being equivalent to a series of site-directed mutagenesis experiments which allows you to see what residues can be changed to what and still maintain protein (or DNA) functionality. They are also one way in which you can get information on the tertiary structure around a residue. Strongly conserved regions are good candidates for active sites, highly variable regions or regions containing insertions and deletions are candidates for loops. The use of multiple sequence alignments for protein structure prediction is discussed in Chapters

16 and 17. There is a large literature on the theory of multiple sequence alignments, much of which can be found on the Web. There is also a large number of programs available on ftp sites for performing multiple sequence alignments on all the most commonly used type of computer equipment. A selection of Web sites that can perform multiple sequence alignment are listed below:

Resource name: ClustalW [6]
Address: http://dot.imgen.bcm.tmc.edu:9331/multi-align/Options/ clustalw.html
Comments: ClustalW is one of the most used multiple sequence alignment packages. It is also freely available on a number of ftp sites for those who wish to download their own copy.

Resource name: CINEMA2
Address: http://www.biochem.ucl.ac.uk/bsm/dbbrowser/CINEMA2/
Comments: a recent version of the Java multiple sequence alignment editor. Nice to use in that it also allows users to edit the alignment and uses color to highlight regions of the alignment where amino acids share similar properties.

Resource name: Match-Box [7,8]
Address: http://www.fundp.ac.be/sciences/biologie/bms/matchbox_ submit.html
Comments: attempts to align proteins on the basis of their amino acid properties, not just sequence. Still an experimental server.

It is important to point out that there is no one best multiple sequence alignment and that the output of the automatic alignment programs can often be improved by human analysis.

14.6 Comparing sequences against a database

Comparing a sequence against a database to discover similarities is one of the most frequently used and most powerful tools in bioinformatics. In essence, it is the same as comparing two sequences, just repeated many thousands of times. However, to perform each pairwise alignment rigorously would be very time consuming, so approximations must be made to allow the searches to run in a reasonable length of time. The two most commonly used programs for comparing an unknown sequence against a sequence database are FASTA [9] and BLAST [10].

FASTA uses a modified form of the Wilbur and Lipman algorithm, making approximations to try and concentrate only on alignments which are likely to be significant. Although FASTA should not miss strong matches, it sometimes misses weak but significant scores. FASTA will attempt to find global alignments.

BLAST (Basic Local Alignment Search Tool) uses an approach based on matching short sequence fragments, and a powerful statistical model to find the best local alignments between the unknown sequence and the database. It is important to note that BLAST will only match continuous sequences – an alignment with insertions and deletions will be displayed as a number of separate fragment matches. A number of programs, for example BEAUTY [11], have been written to make the BLAST output more intuitive.

Both BLAST and FASTA are available in a number of variants useful for different types of sequence matching. These are listed in *Table 14.3*. Because BLAST and FASTA work in different ways, it is usually

TABLE 14.3: *Listing of the BLAST and FASTA programs available for comparing a query sequence against a database*

Program	Probe type	Database type	Comment
BLASTP	p	p	Compares an amino acid query sequence against a protein sequence database
BLASTN	n	n	Compares a nucleotide query sequence against a nucleotide sequence database
BLASTX	n	p	Compares a nucleotide query sequence translated in all reading frames against a protein sequence database
TBLASTN	p	n	Compares a protein query sequence against a nucleotide sequence database dynamically translated in all six frames
TBLASTX	n	n	Compares the six-frame translations of a nucleotide query sequence against a nucleotide sequence database dynamically translated in all six frames
FASTA	p n	p n	Scans a protein or DNA sequence library for similar sequences
TFASTA	p	n	Compares a protein sequence with a DNA sequence library, translating the DNA sequence library on-the-fly
FASTX	n	p	Compares a nucleotide query sequence translated in all reading frames against a protein sequence database

n: nucleic acid sequence or nucleic acid sequence database; p: protein sequence or protein sequence database.

worth running both, one may find significant hits that the other misses and vice versa.

If neither BLAST or FASTA find significant matches, a third and more expensive option is available. A number of sites will allow users to run full Smith–Waterman searches of an unknown sequence against a database using programs such as BLITZ [12]. BLITZ is designed to run on powerful parallel computers, thereby making the complete search more tractable. Although running such programs is expensive in computer time, they can sometimes find weak but significant hits that BLAST and FASTA miss.

Because comparing a sequence against a database is such a fundamental part of bioinformatics, there are many sites which offer BLAST or FASTA database searches. There a number of criteria you should use in deciding which resource to use. Not all the BLAST and FASTA servers offer the same services. From site to site the databases you can search against vary, as do the range of scoring matrices you can use. Also, a number of sites have pages which allow expert users to finely tune their search. In general, all sites allow both DNA and protein database searches to be run against some form of nonredundant sequence database. This database should at least contain all the entries in SWISS-PROT and PIR (for protein) or EMBL and Genbank (for DNA) and is generally a good default choice. Do not be greedy using these resources – for example if you are constructing a contig you only need to run a BLAST or FASTA job for the final assembled sequence, not one for every sequence fragment. Running BLAST against the entire nonredundant database is not an efficient technique for spotting cloning vector contamination.

Resource name: BLAST
Address: http://www.ncbi.nlm.nih.gov/BLAST/
      – the home of BLAST at the NCBI in the US
Address: http://swarmer.stanford.edu/cgi-bin/blastq-
      form?options=simple
      – Stanford, US
Address: http://dot.imgen.bcm.tmc.edu:9331/cgi-bin/seq-
      search/blast_form_local.pl
      – Baylor College, US
Address: http://www.genome.ad.jp/SIT/BLAST.html
      – Japanese genome center
Address: http://genome.eerie.fr/blast/blast-query.html
      – French server
Address: http://www.crihan.fr/www/blast.html
      – another French server

Address: http://ulrec3.unil.ch/software/EPFLBLAST_form.html
            – Swiss server

Some servers require the sequence to be entered in FASTA format. This means that the line of the sequence should start with a title line which must start with a >, followed by the sequence, for example >collagen sequence
gppgppgppgppgppgppgpgpp

Resource name: FASTA
Address: http://www.ebi.ac.uk/htbin/fasta.py?request
            – EBI
Address: http://swarmer.stanford.edu/cgi-bin/fastaq-
            form?options=simple
            – Stanford, US
Address: http://genome.eerie.fr/fasta/fasta-query.html
            – French server
Address: http://www.crihan.fr/www/fasta.html
            – another French server

Resource name: BLITZ
Address: http://www.ebi.ac.uk/searches/blitz.html
Comments: a very sensitive search, but very slow. Should only be used if FASTA and BLAST fail to find a significant match.

Resource name: BEAUTY
Address: http://dot.imgen.bcm.tmc.edu:9331/seq-search/protein-
            search.html
Comments: an enhanced BLAST search which provides extra information on the potential function of an unknown protein sequence. BEAUTY is one of the options available from this general protein sequence search page.

Figure 14.1 shows a typical output from a BLAST search run at the NCBI BLAST server in the US. A C-terminal fragment of collagen X from chick was compared against the nonredundant protein sequence database using default values of the parameters. The server first returns a list of the potentially significant matches. The important numbers to look at are those in the 'High Score' and 'Smallest Sum Probability' columns. For a protein sequence match to be significant, the 'Smallest Sum Probability' should be as small as possible, certainly less than 0.1, and the 'High Score' should be as large as possible. For DNA matches you need to be much more conservative with the scores you use – it is possible to have the 'Smallest Sum Probability' at less than 0.0001 and still have a match that is

```
BLASTP 1.4.9MP [26-March-1996] [Build 14:27:01 Apr  1 1996]

Reference: Altschul, Stephen F., Warren Gish, Webb Miller, Eugene W.
Myers, and David J. Lipman (1990).  Basic local alignment search tool.  J. Mol.
Biol. 215:403-10.

Query= collagenX
       (174 letters)

Database:  Non-redundant GenBank CDS
           translations+PDB+SwissProt+SPupdate+PIR
           246,448 sequences; 69,492,493 total letters.
```

> This number is the probability of the match appearing by chance.

> The alignment score

```
                                                                   Smallest
                                                                     Sum
                                                           High   Probability
Sequences producing High-scoring Segment Pairs:           Score    P(N)      N

sp|P08125|CA1A_CHICK COLLAGEN 1(X) CHAIN PRECURSOR          868   1.7e-116   1
pir||S23297          collagen alpha 1(X) chain precursor ...868   1.7e-116   1
gi|545181            (S68531) type X collagen alpha 1 cha...643   6.1e-89    2
gi|560151            (X65120) alpha1(X)collagen [Homo sap...653   1.4e-86    2
pir||CGHU1D          collagen alpha 1(X) chain precursor ...653   2.3e-86    2
```

> Entries removed for clarity

```
gi|552212            (M32112) major surface antigen p190 ... 59   0.9998    1
pir||PQ0124          major merozoite surface antigen - Pl... 59   0.99992   1
pir||PQ0121          major merozoite surface antigen - Pl... 59   0.99992   1
sp|P29067|ARR2_RAT   BETA-ARRESTIN 2 /gi|203104 (M91590) ... 60   0.99992   1
gi|974218            (U22666) T cell receptor gamma [Gall... 54   0.99992   2
gi|160422            (M55001) merozoite surface antigen 1... 59   0.99992   1
```

> Identity of the matching sequence in the format database|accession number|identifier

> The number of matching fragments between the probe sequence and the entry

```
pir||CGHU1D collagen alpha 1(X) chain precursor - human gnl|PID|e249638
            (X98568) type X collagen [Homo sapiens]
            Length = 680

 Score = 653 (300.4 bits), Expect = 2.3e-86, Sum P(2) = 2.3e-86
 Identities = 119/137 (86%), Positives = 126/137 (91%)

Query:     36 ANQALTGMPVSAFTVILSKAYPGATVPIKFDKILYNRQQHYDPRTGIFTCRIPGLYYFSY 95
              ANQ +TGMPVSAFTVILSKAYP   PI FDKILYNRQQHYDPRTGIFTC+IPG+YYFSY
Sbjct:    542 ANQGVTGMPVSAFTVILSKAYPAIGTPIPFDKILYNRQQHYDPRTGIFTCQIPGIYYFSY 601

Query:     96 HVHAKGTNVWVALYKNGSPVMYTYDEYQKGYLDQASGSAVIDLMENDQVWLQLPNSESNG 155
              HVH KGT+VWV LYKNG+PVMYTYDEY KGYLDQASGSA+IDL ENDQVWLQLPN+ESNG
Sbjct:    602 HVHVKGTHVWVGLYKNGTPVMYTYDEYTKGYLDQASGSAIIDLTENDQVWLQLPNAESNG 661

Query:    156 LYSSEYVHSSFSGFLFA 172
              LYSSEYVHSSFSGFL A
Sbjct:    662 LYSSEYVHSSFSGFLVA 678

 Score = 39 (17.9 bits), Expect = 2.3e-86, Sum P(2) = 2.3e-86
 Identities = 6/14 (42%), Positives = 10/14 (71%)

Query:     11 QSTIPEGYVKGESR 24
              Q+ +PEG++K   R
Sbjct:    518 QAVMPEGFIKAGQR 531

 Score = 39 (17.9 bits), Expect = 2.3e-86, Sum P(2) = 2.3e-86
 Identities = 7/15 (46%), Positives = 11/15 (73%)

Query:     31 FMKAGANQALTGMPV 45
              F+KAG   +L+G P+
Sbjct:    525 FIKAGQRPSLSGTPL 539
```

> A typical alignment between the query sequence and an entry in the database. The middle line shows the residues in common. + signs refer to positions where the nature of the residue is conserved

FIGURE 14.1: *The results of a BLAST search run with the C-terminal of chicken collagen X. The text in the highlighted boxes explains how the Blast output should be interpreted.*

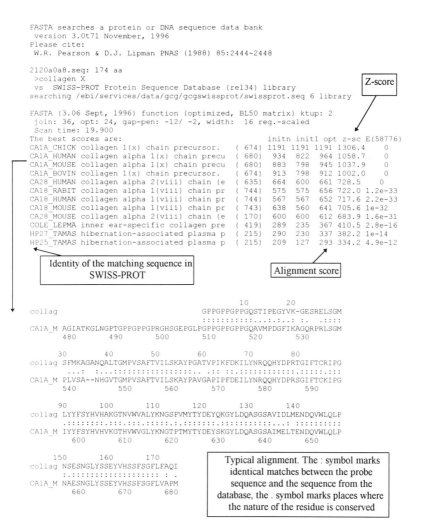

FIGURE 14.2: *Output from a FASTA search of the C-terminal fragment from chicken collagen X compared against the SWISS-PROT database. The text in the boxes shows the positions of the key features of the output.*

essentially random. *Figure 14.2* shows a typical output from a FASTA search run using the FASTA server at the EBI in the UK. The significant statistic from the FASTA output is the z-score. For a significant hit the z-score should be as large as possible. The other useful statistic quoted is the alignment score – again the bigger this number the better.

14.7 When is a hit significant?

It is important to know whether a hit returned by a database search is likely to be biologically significant or not. This question has been investigated extensively. The conclusion for protein global alignments is that any sequences which show 25% (or greater) sequence identity over a stretch of at least 80 amino acids should have the same basic fold (see refs 13 and 14 for a more detailed study). Below this level of similarity we enter what is known as the twilight zone: it is perfectly possible for two sequences to have a similar shape and function with less than 25% identity, it is also possible that they are completely dissimilar proteins – it is impossible to tell from the alignment. The situation for DNA matches is more complex. Because of the redundancy in the DNA code, protein-coding DNA should be translated to protein before performing the search, either manually or using a program such as BLASTX. If the DNA is not from a coding region, then the level of identity needed before the match is biologically significant is not really known. Experience suggests that at least 75% sequence identity should be seen before any given match can be suspected of being significant. However, these statistics will not provide much help for many borderline matches, those with scores which might or might not be significant. In such cases there is no substitute for biological experience and a good knowledge of your system for identifying potentially interesting hits.

References

1. Dayhoff, M.O., Schwartz, R.M. and Orcutt, B.C. (1978) A model of evolutionary change in proteins, in *Atlas of Protein Sequence and Structure*, 5 suppl. National Biomedical Research Foundation, Washington, D.C., Vol. 3, pp. 345–352.
2. Henikoff, S. and Henikoff, J.G. (1992) Amino-acid substitution matrices from protein blocks. *Proc. Natl Acad. Sci. USA*, **89**, 10915–10919.
3. Needleman, S.B. and Wunsch, C.D. (1970) A general method applicable to the search for similarities in the amino acid sequence of two proteins *J. Mol. Biol.*, **48**, 443–453.
4. Smith, T.F. and Waterman, M.S. (1981) Comparison of bio-sequences. *Adv. Appl. Math.* **2**, 482–489.
5. Parry-Smith, D.J., Payne, A.W.R., Michie, A.D. and Attwood, T.K. (1997) CINEMA – A novel Colour INteractive Editor for Multiple Alignments. *GENECOMBIS*, in press.
6. Thompson, J.D., Higgins, D.G. and Gibson, T.J. (1994) CLUSTAL W: improving the sensitivity of progressive multiple sequence alignment through sequence weighting, positions-specific gap penalties and weight matrix choice. *Nucleic Acids Res.*, **22**, 4673–4680.

7. Depiereux, E. and Feytmans, E. (1991) Simultaneous and multivariate alignment of protein sequences: correspondance between physicochemical profiles and structurally conserved regions (SCR). *Protein Eng.*, **4**, 603–613.

8. Depiereux, E. and Feytmans, E. (1992) MATCH-BOX – A fundamentally new algorithm for the simultaneous alignment of several protein sequences. *Comput. Appl. Biosci.*, **8**, 501–509.

9. Lipman, D.J. and Pearson, W.B. (1985) Rapid and sensitive protein similarity searches. *Science*, **227**, 1435–1441.

10. Altschul, S.F., Gish, W., Miller, W., Myers, E.W. and Lipman, D.J. (1990) Basic local alignment search tool. *J. Mol. Biol.*, **215**, 403–410.

11. Worley, K.C., Wiese, B.A. and Smith, R.F. (1995) BEAUTY: an enhanced BLAST-based search tool that integrates multiple biological information resource into sequence similarity search results. *Genome Res.*, **5**, 173–184.

12. Sturrock, S.S. and Collins, J.F. (1993) *MPsrch Version 1.3*. Biocomputing Research Unit, University of Edinburgh, UK.

13. Altschul, S.F., Boguski, M.S., Gish, W. and Wootton, J.C. (1994) Issues in searching molecular sequence databases, *Nature Genet.*, **6**, 119–129.

14. Argos, P., Vingron, M. and Vogt, G. (1991) Protein sequence comparison: methods and significance *Protein Eng.*, **4**, 375-383.

15 Sequencing Projects and Contig Analysis

Andy Brass, University of Manchester, UK

15.1 Introduction

A sequencing project naturally falls into two main parts:

- sequencing and data analysis of the individual clones;
- looking for overlapping clones to assemble into larger contiguous sequence.

The computer needs for these stages are rather different and are therefore discussed separately. In general, the work on clone analysis can be performed readily on the Web. However, as yet there are no generally available Web tools for contig assembly, primarily because of the amount of data that needs to be stored over time. A number of software packages will be discussed which allow users to perform contig assembly, either on a PC/Mac or on a workstation.

15.2 Analyzing clones

Typically, a sequencing reaction will generate about 300–500 bp of data on a given clone. This sequence data is in the form of an autoradiogram, an electropherogram or the equivalent. This can be entered by hand, or using a gel-scanner or other automated gel reader. Automated sequencers have base-calling software which produces a text file. The range of methods for reading sequence into a computer is beyond the scope of this book, but bear in mind that no method is 100% accurate, so the sequence should be checked where possible. We are therefore assuming that all the required sequence information is available in computer readable files. The most important analysis task at this stage is to check the quality of the sequence data. There are two main types of sequencing error that need to be caught at this point:

- the clone contains sequencing vector;
- genetic material from the host organism in which in the clone was propagated has contaminated the clone.

15.2.1 Removing the sequencing vector

A number of databases exist which contain comprehensive lists of sequencing vectors, for example vector-ig [1]. The clone can be compared against these databases using the BLAST tool at http://www.ncbi.nlm.nih.gov/BLAST/. Any good matches found provide clear evidence that the sequence of the contig still contains the sequencing vector, which should be removed before analysis continues. Also, many of the good fragment assembly programs allow users to enter the sequence of the cloning vector into the assembly program so that any matches to the vector can be automatically found and removed (see below). Although this is such a straightforward procedure, the amount of sequence data in the databases which still contains cloning vector is depressing.

Database name: vector-ig
Address: ftp://ncbi.nlm.nih.gov/repository/vector-ig

15.2.2 Removing other cloning sequence artifacts

Genomic sequence from the host organism in which the cloning has been done can enter a clone in two ways:

- the clone is in fact derived entirely from the host;
- the clone is a chimera containing some of the desired sequence plus other material which has jumped into the clone during the cloning process.

Both these eventualities can be tested by using a program such as BLASTN to compare the clone sequence against the entire GenBank or EMBL databases. Clearly any sequence showing a top match with *E. coli* or *S. cerevisiae* sequences needs to be treated with a good deal of caution.

Organisms such as *E. coli* contain mobile genetic elements such as insertion sequences. Such sequences can jump into clones during the sequencing project, indeed this is exactly how some elements (such as IS186 [1]) were first found and identified. The BLASTN output should therefore also be checked for evidence of such mobile elements, or fragments of them, in the clone. Any clones containing such sequences must be treated as suspect.

15.3 Contig assembly

As explained above, there are no programs for contig assembly that can be accessed directly via the Web. However, there are a number of freely available programs that can be downloaded directly from ftp sites which can be used for this purpose (*Table 12.5*). In addition, most commercial software bioinformatics packages will help to manage a sequencing project. To illustrate the process consider the tools available for contig assembly in GCG.

The GCG programs for fragment/contig assembly are:

gelstart:	This program creates a new database for a sequencing project
gelenter:	Allows clone sequences to be entered into the database
gelmerge:	Automatically attempts to find overlaps between clones and construct contigs
gelassemble:	Allows manual editing of the contigs and conflict resolution
gelview:	Displays bar diagrams of the overlap clones in a contig
geldisassemble:	Breaks up the elements in a contig to return them back as single clones

Other software packages offer similar functionality. Probably the best known of the alternative packages is the Staden suite of programs (indeed the GCG programs specifically acknowledge Roger Staden's pioneering work in this area as the inspiration for their programs).

Program name: Staden [2]
Address: http://www.mrc-lmb.cam.ac.uk/pubseq/
Comments: home page of the Staden program, including the program manual and ordering information.

15.4 Predicting protein-coding regions

15.4.1 Coding regions in cDNA

Identification of protein-coding regions in cDNA is relatively straightforward. If the sequencing has been done accurately, a full-length cDNA should contain a single ORF. Partial cDNAs, for example ESTs, are normally sequenced from the 3′ end and should contain the 3′-untranslated region and possibly the C-terminal end of the protein-coding region. Typically you are therefore either looking for a large ORF with appropriate start and stop codons, or an ORF at the start of the sequence with a stop leading into a noncoding region, perhaps with a poly(A) tail. Such signals can most easily be found by translating the DNA into protein in all reading frames and looking for the largest ORF.

Resource: DNA to protein translation
Address: http://expasy.hcuge.ch/www/dna.html
Comments: takes a contributed sequence and translates it into protein – assumes that the ORF will start with the first base in the contributed sequence.

Resource: ORF finder
Address: http://www.ncbi.nlm.nih.gov/gorf/gorf.html
Comments: looks for ORFs in all six frames of the contributed sequence and can then display the ORFs it identifies. A very easy to use Web resource with many nice features. Assumes that the sequence submitted will be in FASTA format. This means that your sequence should start with a title line which begins with a > symbol, e.g.
>my sequence
ggggcacgcatcgactgactgcgcagcatgacatagc

Analysis can be complicated if there are errors in the sequencing (a 1–3% error rate for raw sequence data would be typical). Problems which can occur include:

- frameshift errors or incorrect stop codons leading to abnormally short predicted proteins;
- identification of the correct start codon if there are several candidates near the 5′ end of the cDNA.

The easiest way to spot whether such errors may have occurred is to use a program such as BLASTX which compares ORFs in all six possible frames against protein databases. Frameshifts will appear as matches against similar target proteins in different frames, anomalous stop codons will give matching fragments in the same reading frame. A resource is also available on the Web to try and find frameshift errors in clones. An embarrassing number of published protein sequences have been found to contain such mistakes which could have been spotted had some care been taken with identifying the complete coding region.

Matches against homologous proteins can also help identify the start codon. However, in general, unless you have other experimental evidence available, the best you can normally do is choose the first start in the sequence and note that the sequence data may be ambiguous.

Program name: Frameshift [3]
Address: http://ir2lcb.cnrs-mrs.fr/d_fsed/fsed.html

In general, the usual test of whether you have identified the correct reading frame is by comparison with homologous proteins already in the databases. However, for users with access to the GCG suite of programs, programs such as CODONPREFERENCE allow you to check that the codon usage in the predicted coding region is reasonable for the particular organism from which the DNA originally came. As yet, there are no tools available on the Web that perform a similar function.

Once the protein-coding region has been identified, a program such as TRANSLATE in GCG or the http://expasy.hcuge.ch/www/dna.html Web site can be used to get a protein translation which can be stored in a separate file. Further analysis of the cDNA to identify the encoded protein works best with the protein sequence rather than the DNA. Essentially this is because of the redundancy in the DNA code (third base wobble) – the bases in two DNA sequences might show 67% identity in an alignment but be 100% identical at the protein level. In general, database searches performed using protein sequences can reliably find significant matches to other proteins that show only 25% identity; an equivalent search using DNA can only reliably spot proteins that are more than 40% identical.

15.4.2 Coding regions in genomic DNA

Predicting coding regions in genomic DNA is much more difficult than predicting coding regions in cDNA, particularly for higher eukaryotes. The problem arises with the relative sizes of introns and exons, and can often be a case of looking for a needle in a haystack. Exons of a few hundred bases may well be separated by kilobases of intron. The problem then for the software is to identify all the exons correctly at their proper length without either missing any or predicting false ones, either type of mistake leads to an incorrect prediction of the coding region. For a task of this complexity, no one approach or algorithm seems to work well. The most successful prediction software typically attempts to predict exons in a number of different ways and then generates a consensus prediction from the full set of results. Probably the best known software of this type is the GRAIL suite of programs.

Resource name: GRAIL [4]
Address: http://avalon.epm.ornl.gov/Grail-bin/EmptyGrailForm
Comments: a good general purpose tool for identifying coding regions in DNA sequences.

15.5 DNA analysis

15.5.1 Restriction enzyme maps

The key resource for producing restriction enzymes maps is REBASE, the Restriction Enzyme Database. REBASE is a collection of information about restriction enzymes, methylases, the micro-organisms from which they have been isolated, recognition sequences, cleavage sites, methylation specificity, the commercial availability of the enzymes and references – both published and unpublished observations.

Database name: REBASE
Address: www.gdb.org/Dan/rebase/rebase.html

Most of the programs that calculate restriction enzymes maps access REBASE for information on the types of restriction enzymes and the sites at which they cut. A number of Web sites of resources allow you to paste in a sequence and calculate where either all the different restriction enzymes would cut, or just a subset chosen by the user.

Resource name: WebGene
Address: http://www.bio.indiana.edu/~tjyin/WebGene/RE.html

Resource name: Webcutter
Address: http://www.medkem.gu.se/cutter/

Resource name: Webcutter2
Address: http://www.ccsi.com/firstmarket/firstmarket/cutter/cut2.html
Comment: An updated version of Webcutter – at the time of writing is still in pre-release form.

15.5.2 Promoters and other DNA control sites

There are a number of databases that have collected together information on promoter sites and transcription factor-binding sites. Most of these databases are available from the anonymous ftp site at the EBI.

Database name: EPD (Eukaryotic Promoter Database) [5]
Address: ftp://ftp.ebi.ac.uk/pub/databases/epd/
Comments: in order to be included in EPD, a promoter must be:

* recognized by eukaryotic RNA polymerase II;
* active in a higher eukaryote;
* experimentally defined, or homologous and sufficiently similar to an experimentally defined promoter;
* biologically functional;
* available in the current EMBL release;
* distinct from other promoters in the database.

Database name: TRANSFAC [6]
Address: ftp://ftp.ebi.ac.uk/pub/databases/transfac/
Comments: TRANSFAC is a database on eukaryotic *cis*-acting regulatory DNA elements and *trans*-acting factors. It covers the whole range from yeast to human.

Database name: TransTerm [7]
Address: ftp://ftp.ebi.ac.uk/pub/databases/transterm/
Comments: translation termination signal database, a database of termination and initiation codon contexts.

There is also software available to search for the patterns identified in the databases mentioned above in a given DNA sequence. Programs such as GRAIL (see above) routinely attempt to identify signals such as promoter sequences. In addition, there are number of specialist programs designed to search as sensitively as possible for such patterns.

Resource name: PROMOTER SCAN [8]
Address: http://biosci.umn.edu/software/proscan/promoterscan.htm
Comments: predicts RNA polymerase II promoter sequences by looking for transcription factor-binding sites.

Resource name: SIGNAL SCAN [9]
Address: http://bimas.dcrt.nih.gov:80/molbio/signal/
Comments: finds homologies of published signal sequences in your sequence, mainly transcriptional elements. Most signal elements found probably will not have any meaning, as the elements are in the wrong milieu, wrong cell type or wrong organism. Consequently, there will be many more erroneous signals found by SIGNAL SCAN than significant ones.

Resource name: TFSEARCH
Address: http://www.genome.ad.jp/SIT/TFSEARCH.html
Comments: sensitive searches for transcription factor-binding sites.

15.5.3 RNA secondary structure prediction

A significant amount of work has gone into developing techniques for predicting RNA secondary structure. A number of well-known programs have been developed including Michael Zucker's MFOLD program and the Vienna suite of RNA structure-prediction programs. MFOLD is available on the GCG package and the Vienna programs can be obtained via anonymous ftp. However, there are no secondary structure programs that can be accessed directly from the Web. A PC version of MFOLD is available at ftp://fly.bio.indiana.edu/molbio/ibmpc/pcfldsrc.uue and the Vienna package (for UNIX workstations) can be found at ftp://ftp.itc.univie.ac.at

References

1. Chong, P., Hui, I., Loo, T. and Gillam, S. (1985) Structural analysis of a new GC-specific insertion element IS186. *FEBS Lett.*, **192**, 47–52.
2. Staden, R. (1996) The Staden sequence analysis package. *Mol. Biotechnol.*, **5**, 233–241.
3. Fichant, G. and Quentin, Y. (1995) A frameshift error detection algorithm for DNA sequencing projects. *Nucleic Acids Res.*, **23**, 2900–2908.
4. Xu, Y., Mural, R.J., Einstein, J.R., Shah, M.B. and Uberbacher, E.C. (1996) Grail – a multiagent neural-network system for gene identification *Proc. IEEE*, **84**, 1544–1552.
5. Bucher, P. and Trifonov, E.N. (1986) Compilation and analysis of eukaryotic POL II promoter sequences. *Nucleic Acids Res.*, **14**, 10009–10026.
6. Wingender, E. (1994) The TRANSFAC database. *J. Biotechnol.*, **35**, 273–280.
7. Brown, C.M., Dalphin, M.E., Stockwell, P.A. and Tate, W.P. (1997) The translation signal database, Transterm: Organisms, complete genomes. *Nucleic Acids Res.*, **25**, 246–247.
8. Prestridge, D.S. (1995) Prediction of Pol II Promoter Sequences using Transcription Factor Binding Sites. *J. Mol. Biol.*, **249**, 923–932.
9. Prestridge, D.S. (1991) SIGNAL SCAN: a computer program that scans DNA sequences for eukaryotic transcriptional elements. *CABIOS*, **7**, 203–206.

16 Protein Function Prediction

Andy Brass, University of Manchester, UK

16.1 Introduction

If the contig contains a protein-coding region, the next stage in the analysis is to attempt to determine the function of the expressed protein. There are a number of properties of proteins that can be calculated directly from sequence, for example hydrophobicity, which can be used to predict whether a sequence contains transmembrane helices or a leader sequence. However, in general, the only way we can determine function from sequence is to ask whether the expressed protein is similar to other proteins for which we already have functional information. This can be done in two main ways:

(1) we can ask what known proteins have a sequence that is similar to that of the expressed protein;
(2) we can ask whether the expressed protein contains any subsequences or patterns of conserved residues that are associated with particular protein families or functions.

Figure 16.1 gives an overview of the processes involved in predicting protein function from sequence.

16.2 Comparing a protein sequence against a sequence database to determine function

Proteins with similar sequences should have similar function. Therefore, the most reliable method for determining protein function is to do a database search. The methods of doing this are discussed in

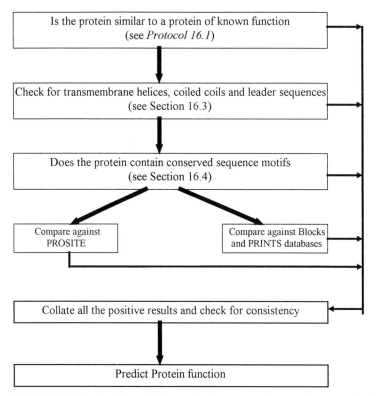

FIGURE 16.1: *Outline of the techniques used to predict protein function from sequence. Because a range of techniques are being used, not all the results generated will typically be consistent with a single functional hypothesis. The most weight should always be placed with the results obtained from the similarity searches, the least weight should be attached to the results from PROSITE.*

Chapter 14. Remember, to be significant, a match should have at least 25% sequence identity over a region of at least 80 amino acids.

There are a number of different tools available for database searching, some slow and sensitive, others faster and more insensitive. The fastest tools (such as BLASTP) should find strong matches very easily and so there is no need to run the more expensive searches (such as BLITZ). These latter tools are only used if programs such as BLASTP do not find significant results. (See *Protocol 16.1.*)

Remember also the importance of scoring matrices. Using a range of scoring matrices is important for a number of reasons. Firstly, the scoring matrix used should be appropriate for the level of match being found: PAM250 for distant matches (~ 25% identity); PAM40 for less closely related proteins; and BLOSUM62 for general use. Secondly, by

PROTOCOL 16.1: Comparing an unknown sequence against a database to determine function

1.   Connect to your local Web site running BLASTP (see Chapter 13 for appropriate sites).

2.   Paste in your protein sequence and run BLASTP choosing BLOSUM62 as the scoring matrix. If you are using the BLAST searches at NCBI, the sequence must be pasted into the box in FASTA format. What that means is the first line should be an identifier starting with a > symbol. For example, if you wanted to search using the sequence:

     dfgerefghalivm . . . . .

     you would paste

     >myseq 1

     dfgerefghalivm . . . . .

     into the sequence box.

     Other BLAST services simply require the raw sequence, i.e. do not include the line starting with >.

3.   Does BLASTP find a good match (see Chapter 13 for details of how to interpret the blast score)? If there is a good match, then does it tell you something about the function of your protein? There can be cases where there is a wonderful match, but to a protein of unknown function.

4.   If BLASTP does not provide clues as to sequence function, try running FASTA. Although FASTA is slower than BLASTP to run, it sometimes finds significant hits that BLASTP misses.

5.   If neither FASTA or BLASTP provide any clues as to protein function, the last program to try is one that performs a full Smith–Waterman search against a database, for example BLITZ at the EBI (http://www.ebi.ac.uk/searches/blitz.html). Both BLASTP and FASTA make approximations to try and make the search more efficient. Programs such as BLITZ make no approximations on the original Smith–Waterman algorithm. Because of this, they are computationally expensive to run, but in return they are very sensitive. Typically, programs such as BLITZ are good at findings matches where the sequence identity is low (~ 20–25%) but over several hundred residues. Such matches are significant but can be missed by programs which make approximations.

using a variety of scoring matrices, it is possible to see which of the borderline matches consistently appear – it is one way of reducing the noise.

In addition to varying the scoring matrix in the database searches, it is also possible to vary the database which you search with the sequence. Typically the databases available to search are a nonredundant protein sequence database, SWISS-PROT and the sequences in the PDB. Certain sites also allow users to search other databases, for example you can search using BLASTP against the OWL composite protein sequence database by using http://www.biochem.ucl.ac.uk/bsm/dbbrowser/OWL/owl_blast.html

16.3 Hydrophobicity, transmembrane helices, leader sequences and sorting

There are a number of functional features that can be predicted directly from protein sequence. For example, hydrophobicity profiles can be used to predict transmembrane helices. There are also a number of small sequence motifs which cells use to target proteins to particular cell compartments (the best known being the KDEL sequence found at the protein C terminus which targets a protein to the endoplasmic reticulum). A number of resources are available on the Web which make use of these properties to help predict protein function from sequence.

16.3.1 Calculating hydrophobicity profiles

Hydrophobicity profiles can be generated and displayed graphically using the ProtScale utility at ExPASy (http://expasy.hcuge.ch/cgi-bin/protscale.pl). This is a very useful utility that allows you to calculate over 50 different properties of proteins where a number (such as hydrophobicity) is assigned to every amino acid. Input to the program can either be a sequence pasted into the sequence window or a SWISS-PROT accession code. The only other parameter needed is the size of the window. The window parameter tells the system how much of a running average to calculate when displaying a value for a specific residue. For example, with a window size of 9, the value plotted for the hydrophobicity of residue number n is actually the average value of the hydrophobicity of residues $n-4$ through to $n+4$. This windowing procedure helps smooth the data and makes hydrophobic and hydrophilic patches stand out more clearly. A typical default would be to set the window size to 9. The only difference from this would be if you were looking for transmembrane helices. Because a transmembrane helix is typically 20 amino acids long, the window size should also be set to 20. *Figure 16.2* shows a typical output from this program.

16.3.2 Predicting transmembrane helices

There are a number of different ways of predicting transmembrane helices in sequences, the simplest being merely to look for regions of the protein containing a run of 20 hydrophobic residues. However, there are also a number of more sophisticated, and accurate, algorithms which can be used not only to predict the location of

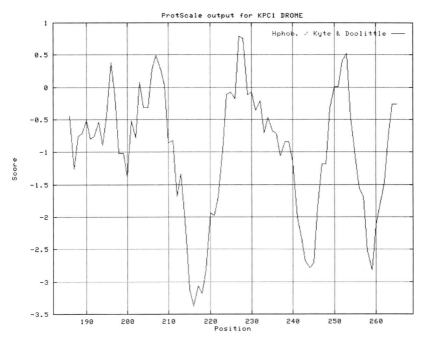

FIGURE 16.2: *The hydrophobicity profile for KPC1_DROME (a protein kinase C from the brain of* Drosophila melanogaster) *calculated using the ProtScale program at http://expasy.hcuge.ch/cgi-bin/protscale.p1*

transmembrane helices but also their orientation in the membrane. Most of these techniques rely on studying the properties of a set of known transmembrane helices. A database of naturally occurring transmembrane helices, TMbase [1], is available by anonymous ftp from ulrec3.unil.ch in the directory pub/tmbase. The Tmbase manual can be found at http://ulrec3.unil.ch/tmbase/TMBASE_doc.html

The available programs are:

Program name: TMPRED
Address: http://ulrec3.unil.ch/software/TMPRED_form.html
Comments: predictions based on a statistical analysis of the TMbase database.

Program name: PHDhtm [2]
Address: www.embl-heidelberg.de/Services/sander/predictprotein/ predictprotein.html
Comments: another of the programs available within the powerful PHD suite of programs.

Program name: TMAP [3]
Address: http://www.embl-heidelberg.de/tmap/tmap_sin.html

Program name: MEMSAT [4]
Anonymous ftp site: ftp.biochem.ucl.ac.uk
ftp directory: /pub/MEMSAT
Comments: not available on the Web: available as a program for IBM-compatible PCs.

These programs all use different statistical models to predict transmembrane helices. In general, they are around 80–95% accurate in predicting the location of helices and their orientation in the membrane. Transmembrane helices are one of the features of a protein that can be predicted with a fair degree of certainty from sequence data.

16.3.3 Leader sequences and protein localization

Proteins can contain signals within their sequence which assist in their processing within the cell, for example leader sequences or signals which target proteins to specific compartments within cells. Web resources are available to help predict both these types of sites.

Program name: SignalP [5]
Address: http://www.cbs.dtu.dk/services/SignalP/
Comments: predicts leader sequences and cleavage sites within both prokaryotes and eukaryotes.

Program name: PSORT [6]
Address: http://psort.nibb.ac.jp/form.html
Comments: analyzes prokaryotic or eukaryotic sequences and searches for protein sorting signals. The program reports back a probability for the protein being localized to different compartments within the cell. The accuracy of this program is reported to be around 60% for eukaryotic sequences, and slightly higher for prokaryotic sequences.

16.3.4 Coiled-coils

Another functional motif that can be identified readily from sequence is the coiled-coil arrangement of α-helices. In this structure, two helices wind round each other to form a very stable structure held together at their interface by hydrophobic interactions. Coiled-coils can be found in many proteins, for example leucine zippers in transcription factors, myosins, etc. A Web page dedicated to coiled-coil proteins can be found at http://www.york.ac.uk/depts/biol/Web/coils/coilcoil.html

Two algorithms are available on the Web for predicting coiled-coil regions:

Program name: COILS [7]
Address: http://ulrec3.unil.ch/software/COILS_form.html
Comments: the original Lupas coiled-coil prediction software.

Program name: Paircoil [8]
Address: http://ostrich.lcs.mit.edu/cgi-bin/score
Comments: a relatively new algorithm, which is supposedly more stringent than the COILS program.

16.4 Comparing a protein sequence against motif and profile databases to determine function

Often, the sequence of a protein is too distantly related to any in the database to allow a reliable identification to be made by sequence alignment. Alternatively, sequence alignment might find a match, but to a protein of no known function. In these cases, there is still a considerable amount that can be done to predict function using bioinformatics tools.

Different regions of a protein evolve at different rates: some parts of a protein must retain a certain pattern of residues for the protein to function. By identifying such conserved regions, it is possible to make predictions about the protein function. For instance, there are many short sequences that are diagnostic of the active site or binding region of proteins, for example integrin receptors recognize either RGD or LDV motifs in their ligands. If the sequence of your protein contains an RGD motif it is possible to predict that one of its functions might be to bind integrins. Such an identification does not mean that the protein *does* bind integrin (there are many examples of RGD-containing proteins which do not bind integrin); however, it does provide an experimentally testable hypothesis as to protein function. Other examples of conserved sequences can be found around the active sites of enzymes, sites of post-translational modification, binding sites for co-factors, protein sorting signals, etc. A number of bioinformatics resources have been developed both to build databases of conserved patterns and to search for instances of such patterns in sequences.

Two main techniques are available for searching for sequence motifs. The first relies on matching patterns of residues using a consensus sequence or motif (covered in Section 16.4.1). This technique has the advantage of being quick to run and the databases of motifs are large and extensive. However, the use of consensus sequences or motifs does have a drawback in that they are somewhat insensitive: only patterns which *exactly* match consensus sequence are reported as hits, sequences which *almost* match are completely ignored. This can be a severe limitation if you are trying to spot complex patterns, for example those found in the helices of G-protein-coupled receptors. In these cases, a second and more sophisticated approach using sequence profiles works better (see Section 16.4.2). In essence, a profile search matches sequences against a full multiple sequence alignment of the conserved region (not just the consensus sequence), and can be much more sensitive at picking out distantly related sequences. However, the creation of profiles and profile databases is a nontrivial task, requiring both computational and human resources. Therefore, the profile databases do not contain as many entries as the motif databases. In practice, sequences should always be searched against both types of databases, significant matches found by one can be completely missed by the other, and vice versa.

16.4.1 Motif databases – PROSITE

The best known motif database is PROSITE, constructed by Amos Bairoch, the latest version of which at the time of writing (version 14.0 released in February, 1997) contains 1167 patterns [9]. A typical entry in PROSITE would be that for the consensus sequence of a casein kinase II phosphorylation site: [ST]-x(2)-[DE], that is a serine or threonine, followed by any two residues, followed by a D or E. The entry in the database also contains significant extra information about the site, what it does, where it has been found, etc.

Database name: PROSITE
Address: http://expasy.hcuge.ch/sprot/prosite.html
Search form: http://expasy.hcuge.ch/sprot/scnpsit1.html
Comments: searching a sequence against the PROSITE database is very straightforward, simply paste the appropriate sequence into the form. Sites allowing PROSITE searches are also available in the US, Australia and Japan (simply use prosite as the keyword in a search of http://www.ebi.ac.uk/htbin/bwurld.pl to find the most convenient site for you).

Many of the patterns in PROSITE (such as the casein kinase site described above), or the *N*-linked glycosylation site (N){P}(S,T){P} (where the {P} notation means any residue except proline) are very small and have a reasonably high chance of appearing at random in proteins. PROSITE massively overpredicts sequences such as these, for example if PROSITE is to be believed it would seem that every other protein contains a myristylation site! Most PROSITE sites therefore give you the option of not including these patterns which give a high rate of false positives in your search. It should always be remembered that PROSITE is a rather blunt sequence analysis instrument. A significant number of the matches it reports will be false-positive. Similarly, many complex patterns can fail to find their match in a test sequence because of a slight variation in the test protein that causes a single mismatch against the consensus sequence. However, PROSITE does contain an extensive collection of patterns and is one of the tools that should always be used in function prediction.

16.4.2 Profile databases

As explained above, profiles provide a more sensitive tool for searching for conserved motifs in proteins. A number of profile databases are available on the Web. As with other bioinformatics resources, there is a trade off between size and quality; some of the resources contain several thousand poorly annotated entries, others contain less than a thousand high quality entries with excellent annotation. An additional issue with profile databases is the quality of the multiple sequence alignment used to generate the profile. The databases with the largest numbers of entries rely on automated alignments, and these can sometimes be nonoptimal. The smaller profile databases have typically invested much time and effort in producing high quality, manually checked alignments. As usual, you will probably need to search all the resources listed to make sure that nothing is missed. A detailed description of these resources and access details are given below. Sometimes the output of these programs is rather complex – the help files give a lot of useful information on how to interpret the significance of the results obtained.

Database name: BLOCKS [10]
Address: http://www.blocks.fhcrc.org/blocks/
Search form: http://www.blocks.fhcrc.org/blocks_search.html
Comments: the BLOCKS database is constructed by producing ungapped multiple alignments for the motifs represented in the PROSITE database. Users also have the ability to create their own BLOCKS. The latest version of the BLOCKS database (9.2) contains 3363 profiles.

Database name: PRINTS database [11]
Address: http://www.biochem.ucl.ac.uk/bsm/dbbrowser/PRINTS/
PRINTS.html
Alignments: http://www.biochem.ucl.ac.uk/bsm/dbbrowser/ALIGN/
ALIGN.html
Search form: http://www.biochem.ucl.ac.uk/cgi-bin/attwood/Search
PrintsForm2.pl
Comments: the PRINTS database has entries corresponding to
conserved regions in multiple sequence alignments. Although it is the
smallest of the databases, it has the most carefully constructed
alignments and the most comprehensive annotation. It is unusual in
that one alignment (fingerprint) can generate a number of profiles, for
example in the alignment of seven-transmembrane-helix proteins
there is one entry for each helix. This improves the sensitivity of the
search – if an unknown sequence contains matches to more than one
motif in the same fingerprint it improves the level of confidence in the
prediction. For example, a protein matching against six of the motifs
for the seven-transmembrane-helix proteins can be predicted more
confidently than a protein that only matches against one. PRINTS
currently contains 600 fingerprints comprised of 3000 motifs (version
13.0). The alignments used to generate the fingerprints can also be
viewed at the Web site (address given above).

Database name: ProDom database [12]
Address: http://protein.toulouse.inra.fr/prodom/prodom.html
Search form: http://protein.toulouse.inra.fr/prodom/blast_form.html
Comments: ProDom currently contains alignments for 9600 protein
domain motifs (version 33). These alignments are generated by an
automatic compilation of homologous domains detected in the SWISS-
PROT database. However the price paid for this extensive coverage is
a relatively low quality annotation. A modified version of the BLAST
program is used to compare unknown sequence against this database.

References

1. Hofmann, K. and Stoffel, M. (1993) TMbase - A database of membrane spanning
   proteins segments. *Biol. Chem. Hoppe-Seyler*, **347**, 166.
2. Rost, B., Casadio, R., Fariselli, P. and Sander, C. (1995) Transmembrane helices
   predicted at 95% accuracy. *Protein Sci.*, **4**, 521–533.
3. Milpetz, F., Argos, P., Persson, B. (1995) TMAP - A new E-mail and WWW service
   for membrane-protein structural predictions. *Trends Biochem. Sci.*, **20**, 204–205.
4. Jones, D.T., Taylor, W.R. and Thornton, J.M. (1994) A model recognition approach
   to the prediction of all-helical membrane protein structure and topology.
   Biochemistry, **33**, 3038–3049.

5. Nielsen, H., Engelbrecht, J., Brunak, S. and von Heijne, G. (1997) Identification of prokaryotic and eukaryotic signal peptides and prediction of their cleavage sites *Protein Eng.*, **10**, 1–6.

6. Nakai, K. and Kanehisa, M. (1992) A knowledge base for predicting protein localization sites in eukaryotic cells. *Genomics*, **14**, 897–911.

7. Lupas, A., Van Dyke, M. and Stock J. (1991) Predicting Coiled Coils from Protein Sequences. *Science*, **252**, 1162–1164.

8. Berger, B., Wilson, D.B., Wolf, E., Tonchev, T., Milla, M. and Kim, P.S. (1995) Predicting Coiled Coils by Use of Pairwise Residue Correlations. *Proc. Natl Acad. Sci. USA*, **92**, 8259–8263.

9. Bairoch, A., Bucher, P. and Hofmann, K. (1997) The PROSITE database, its status in 1997. *Nucleic Acids Res.*, **25**, 217–221.

10. Henikoff, S. and Henikoff, J.G.(1991) Automatic assembly of protein blocks for database searching. *Nucleic Acids Res.*, **19**, 6565–6572.

11. Attwood, T.K., Beck, M.E., Bleasby, A.J. and Parry-Smith, D.J. (1994) PRINTS – a database of protein motif fingerprints. *Nucleic Acids Res.*, **22**, 3590–3596.

12. Sonnhammer, E.L.L. and Kahn, D. (1994) Modular arrangement of proteins as inferred from analysis of homology. *Protein Sci.*, **3**, 482–492.

17 Protein Structure Prediction

Andy Brass, University of Manchester, UK

17.1 Introduction

Protein structure is normally thought of at four levels:

- primary – the sequence;
- secondary – the pattern of α-helices and β-sheets;
- tertiary – the arrangement of the residues in space;
- quaternary – the protein–protein interactions.

In recent years, another level of protein structure lying between secondary and tertiary structure – known as the protein fold – has proved to be very useful [1]. A 'fold' describes the way in which the secondary structure elements in a protein are packed together, without giving full detail of all the loop structures or the exact coordinates of the residue. The concept of folds is demonstrated in *Figure 17.1*.

Techniques for predicting secondary structure from sequences or multiple sequence alignments are well defined and understood. These techniques are summarized in Section 17.3. Predicting the tertiary structure is rather more difficult. It used to be the case that the only way that tertiary structure could be predicted was to ask whether the probe sequence was homologous to a protein of known structure. This could be done by comparing the unknown sequence against a protein sequence database, and a number of database resources were developed to assist in this process (described in Section 17.2). It is still the case that this method gives by far the most accurate prediction of protein tertiary structure (described in Section 17.4.1).

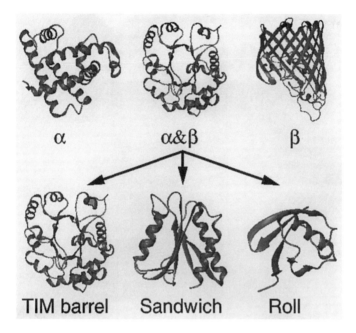

FIGURE 17.1: *The hierarchy of protein structure. Proteins can typically be classified at the highest level of having an α structure, and α&β structure or a β structure. These structures can then be further classified at the level of protein architecture, i.e. the positioning of the secondary structure elements within the protein. For example α&β structures can be divided into TIM barrels, sandwiches or rolls. Examples of proteins containing a sandwich architecture would include flavodoxin and β-lactamase. Adapted from the CATH resource web pages, http://www.biochem.ucl.ac.uk/bsm/cath/lex/cathinfo.html, with permission from the University College London Biomolecular Structure and Modelling Unit.*

However, this technique is not generally applicable because in around 80% of cases the protein whose sequence is known will not be similar to a protein of known structure. However, in recent years, new techniques have become available that allow a structure to be predicted from sequence when there is no similarity to a protein of known structure. It has been shown that there are only a limited number of protein folds found in nature, and that certain folds are actually quote common. A number of new 'threading' algorithms exploit this fact to provide a method of predicting the tertiary structure of a protein (at least at the level of the fold). These programs attempt to 'thread' the unknown sequence onto every structure in a fold library and then ask which of these structures seem most sensible for the sequence in question. Resources for fold recognition and threading are described in Section 17.4.2.

17.2 Protein structure resources

The primary resource for protein structure is PDB. This database stores the coordinates of protein structures that have been solved either using X-rays or NMR. The current release of PDB (January 1997) contains over 4700 protein structure entries, 217 protein/nucleic acid complexes, 386 nucleic acids and 12 carbohydrate entries.

Database name: PDB
Address – US: http://www.pdb.bnl.gov/
Address – Europe: http://www2.ebi.ac.uk/pdb/
Comments: at the moment the PDB is the major repository of protein structure information. This is likely to change in the near future as a replacement database is created. However, it is not yet clear what the name and form of the replacement service will be.

A number of software tools are available which enable PDB entries to be viewed on a PC or workstation. Perhaps the most popular of these programs is rasmol. Instructions for downloading and installing rasmol are given in *Protocol 12.1.*

Web resources are also available which help classify the data from PDB in ways which make it more usable than the raw coordinate data. Some resources concentrate on providing sequence data from the protein entries, others classify the proteins into the structural families based on their three-dimensional similarities. These latter resources are particularly valuable to help interpret results from fold recognition programs.

Database name: NRL-3D
Address: http://www.gdb.org/Dan/proteins/nrl3d.html
Comments: a database of the sequences of all the proteins whose structure has been solved. It is possible to use NRL-3D as the target database for sequence similarity searches and therefore discover whether a given sequence is related to one of known structure.

Database name: HSSP (homology-derived structures of proteins) [2]
Address: http://www.sander.embl-heidelberg.de/hssp/
Comments: a multiple sequence alignment has been carried out for each protein of known three-dimensional structure from PDB and stored on the database. The listed homologs are very likely to have the same three-dimensional structure as the PDB protein on which they have been aligned. HSSP therefore gives a listing of all proteins in

SWISS-PROT for which a structure can be relatively confidently defined by homology.

Database name: SCOP (structural classification of proteins) [3]
Address: http://scop.mrc-lmb.cam.ac.uk/scop/
Comments: this resource hierarchically classifies proteins of known structure (the classification scheme used seems to work very well). It allows you to ask what other proteins are structurally similar to a given protein of known structure. The site at Cambridge also allows BLAST searches to be run against a database of the sequences of known structures using the address http://scop.mrc-lmb.cam.ac.uk/scop/aln.cgi

Database name: CATH [4]
Address: http://www.biochem.ucl.ac.uk/bsm/cath/
Comments: CATH is similar to SCOP in that attempts a hierarchical classification of known protein structures.

17.3 Secondary structure prediction

There exists a huge literature on techniques for predicting protein secondary structure from sequence. In essence, however, the papers can be split into two types: those dealing with secondary structure prediction from single sequences, and those dealing with predictions from multiple sequence alignments.

Until recently, secondary structure prediction was regarded with a high degree of suspicion. The majority of the algorithms took as their input a single sequence. Even the best known algorithms (such as Chou – Fasman [5] and GOR [6]) only predicted structures to around 60% accuracy, and there were certain structures – particularly those rich in β-sheet – on which the algorithms failed spectacularly. The reasons for this failure was predominantly that information from a single sequence only tells you about the neighbors of a residue in sequence and not in space. It seems that a significant determinant of local secondary structure is the protein fold itself – this information is missing from just a single sequence.

Two developments changed this state of affairs. Firstly, it was realized that multiple sequence alignments could be used to improve the predictive power of the prediction. A multiple sequence alignment can be thought of as nature's version of a mutagenesis experiment: analyzing the allowed variations at a point in a sequence does provide information on the tertiary structure of the protein at that point. Secondly, neural networks began to be used to develop automatic rules for assigning structure from sequence. It is now accepted that given a good, large multiple sequence alignment it is possible to predict protein secondary structure very accurately – typically producing predictions that are 10% more accurate than predictions made from single sequences. Indeed a number of papers have shown that the prediction accuracy now being attained by programs such as PHD (see below) is at the maximum level possible [7,8].

Program name: PHDsec [9]
Address: http://www.embl-heidelberg.de/predictprotein/predictprotein.html
e-mail server: predictprotein@embl-heidelberg.de
Comments: the predict protein server provides a variety of services from secondary structure prediction to fold recognition. There is extensive help documentation available. Although many secondary structure prediction programs are available on the Web, there is a general consensus that this is the best available program.

17.4 Tertiary structure prediction

17.4.1 Comparison against sequences of known structure

There are a number of ways in which sequences can be compared against databases of known structure. The easiest is to use the BLASTP program to compare a sequence against either NRL-3D or the sequences stored at SCOP (see above). If matches greater than 40% identity over more than 100 residues are found then the unknown protein will have a very similar structure to the known protein. In such cases, homology modeling (see below) should prove very successful in predicting a detailed and accurate model of the unknown protein. At between 25 and 40% identity the folds of the two proteins will be the same, but homology modeling becomes more difficult and inaccurate.

If the comparison against NRL-3D does not find a match, the next step is to see if the unknown sequence is related to any of the sequences in the HSSP database (see above). The easiest way to do this is to use BLAST or FASTA to compare the unknown sequence against a protein sequence database (SWISS-PROT, TREMBL or PIR). A resource such as SRS should then be used to retrieve any hits showing greater than 25% sequence identity. If any of these hits are contained within the HSSP database then this will be reported in the DR line of the database annotation (see *Table 13.4*). If the unknown protein does have greater than 25% sequence identity to a protein in HSSP then it should be safe to assume that the unknown protein at least shares a fold with the HSSP protein. At the moment, the number of proteins represented in NRL-3D and HSSP is sufficient to ensure that about 20% of proteins will have a known structural homolog.

17.4.2 Homology modeling

In general, a discussion of homology modeling is beyond the scope of this book and requires specialist molecular modeling and molecular graphics resources to be carried out well. However, there is one resource on the Web that attempts this task automatically, Swiss-Model [10].

Program name: Swiss-Model
Address: http://expasy.hcuge.ch/swissmod/SWISS-MODEL.html
Comments: an automated protein modeling server running at the Geneva Biomedical Research Institute, Glaxo-Wellcome Research and Development S.A., Switzerland. Swiss-Model can work in two ways. If you submit a sequence it will compare it against a database of known structures. If a match is found to a protein of known structure an alignment is carried out. This alignment is then used to thread the unknown sequence onto the known structure. This raw structure prediction is then refined using molecular modeling. The coordinates of the unknown protein are then returned to you, either to be viewed using rasmol or the protein structure viewer developed to work with Swiss-Model (details on the Swiss-Model Web page). Alternatively, users can submit their own alignment of the unknown protein with one of known structure to use as the basis for the homology modeling.

17.4.3 Threading algorithms and fold recognition

One of the big recent growth areas in protein structure has been fold recognition and threading. The excitement about these techniques is that they provide a method of predicting protein structure from sequence in cases where there is no identifiable homology to proteins of known structure.

The basic idea of threading is straightforward. Rather than asking 'out of the whole universe of possible protein shapes, what shape is this sequence likely to adopt?', you ask 'I have observed the folds used by proteins of known structure, is it possible that this sequence could be folded into one of these known structures?' The first question involves searching billions upon billions of possible structures, the second involves searching less than 1000. The chances of success are improved further by the observation that there are certain protein folds that appear over and over again – the majority of newly crystallized proteins will be found to have a fold related to one we have seen before.

In threading, the unknown sequence is threaded onto a fold template from a library in some optimal way, and the energy of the sequence is calculated (again, a large literature has been created on exactly what the best energy functions to use are). After the sequence has been threaded onto all the templates in the library the scores are compared and the software decides whether any of the matches are significant.

Fold recognition is not a particularly reliable technique – it would be safe to assume that the correct fold is only identified between 30 and 50% of the time (which is actually a very impressive achievement given the complexity of the task). The results from these programs are also rather coarse-grained; in most cases they would not be good enough to form the basis of a homology modeling study. However, it is the only technique available for predicting structural information for the vast majority of proteins. A number of resources are available for fold recognition, some on the Web and some as code which can be downloaded onto UNIX workstations. These programs are summarized below. In general, it would seem to be a good idea to use as many of these resources as possible and check to see whether there is any consensus between them on the predicted fold.

Program name: TOPITS [11]
Address: http://www.embl-heidelberg.de/predictprotein/predict protein.html
Comments: another strand to the powerful PHD package available from Chris Sander's group. This package attempts to integrate information from the secondary structure and accessibility predictions from PHD to thread sequences onto templates. It works best where a good multiple alignment can be constructed, but less well in cases where only small alignments can be built.

Program name: frsvr (fold recognition server)
Address: http://www.mbi.ucla.edu/people/frsvr/frsvr.html
Comments: UCLA-DOE structure prediction server. It provides a wide variety of analyses on the submitted sequence and provides a Web page to summarize the results – another good site at which to start an analysis.

Program name: 123D [12]
Address: http://www-lmmb.ncifcrf.gov/~nicka/123D.html
Comments: combines data from secondary structure predictions with residue hydrophobicity to calculate the fit to the different fold templates.

Program name: THREADER and THREADER2 [13]
Web page: http://globin.bio.warwick.ac.uk/~jones/threader.html
Comments: one of the first threading programs. At present, this software is not available to run via a Web server, however the code can be obtained from David Jones by following the instructions given on the Web page. Threader uses single sequences and does not include information from any form of secondary structure prediction. THREADER, and the recent update, THREADER2 have been tested extensively. Results seem to suggest that THREADER2 is probably the most sensitive fold recognition program available at the moment.

Program name: ProFIT [14]
Web page: http://lore.came.sbg.ac.at/Extern/software/Profit/profit.html
Comments: protein fold identification tool. This resource is also not available on a Web server, but the program can be downloaded from an ftp site. Like THREADER, ProFIT uses single sequences and ignores secondary structure predictions.

17.5 Critical assessment of structure prediction (CASP)

There are many techniques now available for protein structure prediction. A problem for researchers is knowing how much to believe each of the prediction methods. To this end, two international protein structure prediction contests have been staged, CASP1 and CASP2. In CASP1, staged in 1994, experimentalists provided information on structures that were about to be solved but for which coordinates were not yet available. The theorists then attempted to predict the structure of these sequences. At the CASP1 meeting, the results of the theory were compared with experiment to provide an objective overview of the predictive power of the methods used. Some 135 predictions were made by 35 different groups. The results are published in a special issue of *Proteins: Structure, Function and Genetics*, **23**, No. 3, November 1995.

This process has been repeated recently at the CARP2 meeting, held in December 1996. Four classes of test were considered:

- Comparative modeling, i.e. homology modeling;
- Fold recognition ('threading'): testing a sequence for compatibility against a library of folds;
- *Ab initio* structure prediction: deriving structures, approximate or otherwise, from sequence;
- Docking – predicting the mode of association of ligands and proteins.

Again, the results of the comparisons will be published, but for the mean time information on CASP2 is available from two Web sites:

Europe: http://www.mrc-cpe.cam.ac.uk/casp2/
US: http://iris4.carb.nist.gov/casp2/

The CASP conferences provide one of the best sources of information on structure prediction protocols, and anyone interested in getting seriously involved in this field should refer to the CASP sites for a wealth of information on good and bad practice.

References

1. Orengo, C.A., Jones, D.T. and Thornton, J.M. (1994) Protein superfamilies and domain superfolds. *Nature*, **372**, 631–634.
2. Sander, C. and Schneider, R. (1993) The HSSP data base of protein structure – sequence alignments. *Nucleic Acids Res.*, **21**, 3105–3109.
3. Murzin, A.G., Brenner, S.E., Hubbard, T. and Chothia, C. (1995) scop: a structural classification of proteins database for the investigation of sequences and structures. *J. Mol. Biol.*, **247**, 536–540.
4. Michie, A.D., Orengo, C.A. and Thornton, J.M. (1996) Analysis of domain structural class using an automated class assignment protocol. *J. Mol. Biol.*, **262**, 168–185.
5. Chou, P.Y. and Fasman, G.D. (1978) Prediction of the secondary structure of proteins from their amino acid sequence. *Adv. Enzymol.*, **47**, 45–148.
6. Garnier, J., Gibrat, J.F. and Robson, B. (1996) GOR method for predicting protein secondary structure from amino-acid sequence. *Methods Enzymol.*, **266**, 540–553.
7. Russell, R.B. and Barton, G.J. (1993) The limits of protein structure prediction accuracy from multiple sequence alignment. *J. Mol. Biol.*, **234**, 951–957.
8. Rost, B., Sander, C. and Schneider, R. (1994) Redefining the goal of protein secondary structure prediction. *J. Mol. Biol.*, 235, 13–26.
9. Rost, B., Sander, C. and Schneider, R. (1994) PHD – an automatic mail server for protein secondary structure prediction. *Comput. Appl. Biosci.*, **10**, 53–60.
10. Peitsch, M.C. (1996) ProMod and Swiss-Model: Internet-based tools for automated comparative protein modelling. *Biochem. Soc. Trans.*, **24**, 274–279.
11. Rost, B. (1995) TOPITS: Threading one-dimensional predictions into three-dimensional structures. In: *The Third International Conference on Intelligent Systems for Molecular Biology (ISMB)* (C. Rawlings, D. Clark, R. Altman, L. Hunter, T. Lengauer and S. Wodak, eds). AAAI Press, pp. 314–321.

12. Alexandrov, N.N., Nussinov, R. and Zimmer, R.M. (1995) Fast protein fold recognition via sequence to structure alignment and contact capacity potentials. In: *Pacific Symposium on Biocomputing '96* (L. Hunter and T.E. Klein, eds). World Scientific Publishing Co., Singapore, pp. 53–72.
13. Jones, D.T., Taylor, W.R. and Thornton, J.M. (1992) A new approach to protein fold recognition, *Nature*, **358**, 86–89.
14. Flockner, H., Braxenthaler, M., Lackner, P., Jaritz, M., Ortner, M. and Sippl, M.J. (1995) Progress in fold recognition. *Proteins: Struct. Funct. Genet.*, **23**, 376–386.

Appendix A

Glossary

Allele: one of the possible different forms of a gene at a given locus.

Amplification: increasing the number of copies of a specific DNA molecule.

Anneal: the hybridization of a single-stranded DNA molecule to another single-stranded DNA molecule of complementary sequence.

Annotation: the information included with the sequence data in an entry in a sequence database.

Autoradiography: the process of exposing X-ray film to a radioactive source to generate an image of the radioactive components of that source. The developed film showing the image is called an autoradiogram or autoradiograph.

Base pair (bp): a pair of complementary nucleotides in double-stranded DNA.

Blunt end: an end of a double-stranded DNA fragment in which neither strand protrudes past the end of the other.

cDNA (complementary DNA): DNA synthesized from (and so complementary to) an RNA template.

Chromosome: a structure of DNA and associated proteins that contains the hereditary information of the cell in the form of a linear array of genes.

Codon: a section of nucleic acid capable of encoding a single amino acid. Codons are three bases long.

Complementary sequence: sequence which will hybridize with a specified sequence. Nucleic acid strands run antiparallel, and 'A' pairs with 'T' and 'G' with 'C', so the complement of 5'-AGCTC-3' is 5'-GAGCT-3'.

Compressions: regions of a sequencing gel where the normally even spacing of bands is disrupted by anomalous migration of specific bands.

Contiguous sequence ('contig'): a long stretch of sequence built up from a number of shorter, overlapping segments of sequence.

Cosmid: a modified plasmid containing sequences from bacteriophage λ that allow the insertion of large DNA fragments. Cosmids are circular double-stranded molecules typically 45–48 kb long, including the vector sequences.

Co-termination: when the polymerase stops elongating the chain without incorporating a ddNTP.

Denature (melt): to separate the two complementary strands of double-stranded DNA. This is normally accomplished by treatment with alkali or by use of elevated temperature.

DNA (deoxyribonucleic acid): nucleic acid comprising the nucleosides deoxyadenosine, deoxycytidine, deoxyguanosine and deoxythymidine.

2′, 3′ -dideoxynucleotide: a 2′-deoxyribonucleotide analog which does not have a hydroxyl group at the 3′ position. Incorporation of such an analog prevents further extension of the strand, resulting in chain termination.

Diploid: a cell having two chromosome sets, or an individual with two chromosome sets in each of its cells. Humans are diploid.

Electrophoresis: a technique for separating molecules based on their differential mobility in an electric field.

Electropherogram: the best-known output from an automated sequencer: a four-color representation of a sequence, showing peaks which represent bases.

Exon: RNAs for most eukaryotic genes are processed after transcription. This processing includes the removal of some sections of the RNA ('intervening sequences' or 'introns'). The sections which remain in the mature (processed) message are known as exons.

Exonuclease: an enzyme that degrades DNA from the end of the molecule.

Fidelity: the accuracy with which a polymerase synthesizes the strand complementary to the template DNA.

Fluorochrome (fluorophore): a fluorochrome is a dye which absorbs light at (approximately) one wavelength and then emits light at another (longer) wavelength. These characteristic wavelengths are the absorption and emission spectra respectively.

Frameshift: when the normal protein coding information changes from one frame to another. This may be due to mutation or a sequencing error, or may be part of a regulatory mechanism.

ftp: the file transfer protocol used to move computer files from one computer to another on the Internet.

Gene: the fundamental physical and functional unit of heredity. Many genes encode proteins.

Genome: the complete set of heritable components of an organism: all its genes.

Genotype: the particular set of alleles present in an individual organism.

Haploid: a cell having one chromosome set, or an individual with one chromosome set in each of its cells.

Hapten: a small molecule which can be detected by its high-affinity binding to a detector molecule, for example biotin can be detected by use of avidin, streptavidin or a specific antibody.

Heterozygote: a (diploid) cell or organism having two different alleles at a given locus.

Homology: the similarity of nucleotide sequence between two distinct DNA molecules. Strictly, this is 'sequence similarity', 'homology' implies an evolutionary relationship between the two molecules.

Homozygote: a (diploid) cell or organism having the same allele at both copies of a given locus.

Hybridization: the process of complementary base pairing between two single strands of nucleic acid.

Internet: the communication protocol used to connect computers on the World Wide Web.

Intron: *see* 'Exon'.

kb (kilobase or kilobase pair): a measure of the length of a nucleic acid molecule. One kilobase is 1000 bases or base pairs of nucleic acid.

Label: a detectable entity covalently linked to a nucleic acid. Typical labels include radioisotopes, fluorochromes and haptens.

Ligase: an enzyme that joins the ends of two nucleic acid molecules to form a single molecule. This linkage is via a phosphodiester bond between the $5'$-PO_4 end of one molecule and the $3'$-OH end of the other.

Ligate: to covalently join two ends of nucleic acids.

Locus: the position on a chromosome where a particular gene or mutation is located.

Melt: see Denature.

mRNA (messenger RNA): RNA molecules which encode proteins.

Mutation, mutant: mutations change one DNA sequence into another by adding, subtracting or altering one or more bases. Mutants are the result: their DNA sequence differs from that of their ancestors prior to the mutation event.

Northern blot: the RNA equivalent of a Southern blot.

Nucleoside: a base covalently linked to a sugar (ribose or deoxyribose).

Nucleotide: a base covalently linked to a sugar (ribose or deoxyribose) which is itself linked to one or more phosphate groups.

Oligonucleotide: several nucleotides joined together to form a short, single-stranded DNA molecule.

ORF (open reading frame): a region of sequence between stop codons.

PCR (polymerase chain reaction): an enzymatic method for amplifying a specific segment of DNA.

Phosphorimaging: an alternative to autoradiography.

Plasmid: an extrachromosomal, circular DNA molecule found in some bacteria. Plasmids are replicated in the cell. Plasmids used in molecular biology carry one or more selectable markers to ensure that they are maintained in the cell from one generation to the next.

Polymerase: an enzyme which synthesizes DNA or RNA, using a single-stranded nucleic acid as a template. DNA polymerases generally require a double-stranded region to extend.

Polymorphism: a variation in the sequence of DNA, so that two or more versions exist.

Primer: an oligonucleotide which anneals to its complementary sequence, thus forming a short double-stranded region. This is extended by a DNA polymerase. Primers are required for the chain termination method and for PCR.

Primer extension: the enzymatic extension of an oligo-nucleotide–template hybrid, incorporating nucleotides complementary to the template.

Primer walking: a strategy for sequencing in which a new sequencing primer is designed based on results of the previous sequencing reaction. This primer is then used to sequence the next unknown section of the template. In this strategy, the sequence is determined in a methodical, stepwise fashion.

Protein fold: a schematic representation of the three-dimensional arrangement of the secondary structure in a protein.

Proofreading: the ability of a polymerase to excise incorrectly incorporated nucleotides and replace them with correct ones. Proofreading polymerases have a 3′–5′ exonuclease activity with which they remove the mismatched base.

Proteinase K: an enzyme which degrades proteins.

Reading frame: the codon sequence that is determined by reading a sequence from a given point. As the genetic code is based on triplets of nucleotides, there are three possible reading frames on each strand (i.e. in each direction).

Recombinant: a DNA molecule generated by ligating heterologous DNA molecules.

Restriction endonuclease (restriction enzyme): an enzyme which recognizes a short, specific DNA sequence and cleaves both strands. Many restriction endonucleases are commercially available. They differ in the sequence recognized and the precise location of the cuts in each strand.

RFLP (restriction fragment length polymorphism): DNA can be cut by restriction endonucleases to give a set of DNA fragments. The characteristic lengths of these fragments can be measured by electrophoresis. A polymorphism may lead to the production of fragments of different size from different versions of the sequence. This length polymorphism is an RFLP.

Reverse transcriptase: an enzyme capable of synthesizing DNA from an RNA template (an RNA-dependent DNA polymerase).

RNA (ribonucleic acid): nucleic acid comprising the nucleosides adenosine, cytidine, guanosine and uridine. It is closely related to DNA. Principal differences are the use of ribose instead of 2′-deoxyribose and uridine in place of thymidine. Some RNA molecules encode proteins, others are components of the protein synthesis machinery. Some viruses have RNA genomes.

Scoring matrix: used in sequence alignment to give a numerical value to each of the matches in the alignment.

Secondary structure (of nucleic acid): the folding of a self-complementary nucleic acid molecule.

Self-complementary: a nucleic acid sequence that, on folding, can produce one or more duplex regions by the formation of complementary base pairs.

Southern blot: DNA that has been electrophoretically separated and immobilized on a solid support (nylon or nitrocellulose). Named after Ed Southern, who devised this technique.

Stop codons: codons which terminate translation. These are UAA, UGA and UAG.

Stringency: conditions that affect the specificity of hybridization between nucleic acid molecules. Low stringency conditions allow strands to form hybrids even when they are not perfectly complementary. Increasing the temperature and decreasing the ionic strength increase stringency.

Tailed primer: a primer complementary to the template at the 3′ end, but containing other sequence, such as a 'universal primer' sequence at the 5′ end.

Template: a nucleic acid molecule used by a polymerase to direct the synthesis of a new nucleic acid molecule of complementary sequence.

T_m **(melting temperature):** the temperature at which the transition from double-stranded to single-stranded DNA is 50% complete.

Transcription: the production of RNA molecules by an RNA polymerase, using a DNA template.

Transformation: the introduction of a DNA molecule into an organism (e.g. a bacterium), so that the DNA molecule can be inherited by subsequent generations.

Translation: the production of protein based on the information encoded in an mRNA molecule.

Universal primer: standard primers which anneal to segments of the M13 genome (or its complement). These sequences have been engineered into many common vectors.

Vector: a DNA molecule that is used to clone DNA segments of interest by allowing their propagation in bacteria or other organisms.

Web: the tools and resources available using the World Wide Web on the Internet.

Wild-type: the normal allele of a gene.

Appendix B

Amino acid and nucleotide codes

TABLE B.1: *IUB/IUPAC codes*

Code	Bases	Derivation	Complements
A, C, G, T	A, C, G, T	see *Figure 1.5*	T, G, C, A
R	A or G	puRine	Y
Y	T or C	pYrimidine	R
W	A or T	Weak	W
S	C or G	Strong	S
M	A or C	aMino	K
K	G or T	Keto	M
B	C, G or T	not A	V
D	A, G or T	not C	H
H	A, C or T	not G	D
V	A, C or G	not T	B
N	A, G, C or T	aNy	N

These single-letter codes are used to represent mixed or ambiguous bases. They are derived from shared features of the bases (third column). Strong and weak refer to the strength of the hydrogen bonds between these pairs (see *Figure 1.7*).

TABLE B.2: *One- and three-letter codes for the common amino acids*

Amino acid	Abbreviation		Amino acid	Abbreviation	
	3-letter	1-letter		3-letter	1-letter
Alanine	Ala	A	Leucine	Leu	L
Arginine	Arg	R	Lysine	Lys	K
Asparagine	Asn	N	Methionine	Met	M
Aspartic acid	Asp	D	Phenylalanine	Phe	F
Cysteine	Cys	C	Proline	Pro	P
Glutamine	Gln	Q	Serine	Ser	S
Glutamic acid	Glu	E	Threonine	Thr	T
Glycine	Gly	G	Tryptophan	Trp	W
Histidine	His	H	Tyrosine	Tyr	Y
Isoleucine	Ile	I	Valine	Val	V

TABLE B.3: *The genetic code*

Second letter

		U	C	A	G	
First letter	U	UUU ⎫ UUC ⎭ Phe UUA ⎫ UUG ⎭ Leu	UCU ⎫ UCC ⎪ UCA ⎬ Ser UCG ⎭	UAU ⎫ UAC ⎭ Tyr UAA STOP UAG STOP	UGU ⎫ UGC ⎭ Cys UGA STOP UGG Trp	U C A G
	C	CUU ⎫ CUC ⎪ CUA ⎬ Leu CUG ⎭	CCU ⎫ CCC ⎪ CCA ⎬ Pro CCG ⎭	CAU ⎫ CAC ⎭ His CAA ⎫ CAG ⎭ G1n	CGU ⎫ CGC ⎪ CGA ⎬ Arg CGG ⎭	U C A G
	A	AUU ⎫ AUC ⎬ Ile AUA ⎭ AUG Met	ACU ⎫ ACC ⎪ ACA ⎬ Thr ACG ⎭	AAU ⎫ AAC ⎭ Asn AAA ⎫ AAG ⎭ Lys	AGU ⎫ AGC ⎭ Ser AGA ⎫ AGG ⎭ Arg	U C A G
	G	GUU ⎫ GUC ⎪ GUA ⎬ Val GUG ⎭	GCU ⎫ GCC ⎪ GCA ⎬ Ala GCG ⎭	GAU ⎫ GAC ⎭ Asp GAA ⎫ GAG ⎭ Glu	GGU ⎫ GGC ⎪ GGA ⎬ Gly GGG ⎭	U C A G

Third letter

Appendix C

Suppliers

Below is a list of the major suppliers of equipment, enzymes, reagents and consumables mentioned in the text.

Manufacturers are continually updating and modifying their product range. This list is, therefore, unavoidably incomplete. Exclusion from the list should not be taken to imply that a manufacturer does not provide particular products or services, neither should inclusion be taken as a recommendation for any given application. When selecting materials and apparatus for performing DNA sequencing, you are advised to refer to any specific instructions or caveats issued by the manufacturer.

Cloning vectors: Amersham, Boehringer Mannheim Biochemicals (BCL in the UK), Clontech Laboratories Inc., Gibco-BRL Life Technologies, Invitrogen Corporation, Promega, Stratagene.

Consumables: Corning Costar, Gilson, Midwest Scientific, Perkin-Elmer Corp., Stratagene, Techne.

Custom oligonucleotide synthesis: Appligene, Clontech Laboratories Inc., Cruachem, Gibco-BRL Life Technologies, MWG-Biotech.

DNA sequencing reagents and/or kits: Amersham, Applied Biosystems, Gibco-BRL Life Technologies.

Enzymes and biochemical reagents: Amersham, Boehringer Mannheim Biochemicals (BCL in the UK), Appligene, Clontech Laboratories Inc., Gibco-BRL Life Technologies, New England Biolabs, Promega, Stratagene.

Gel electrophoresis equipment: Appligene, Bio-Rad, FMC, GATC, Gibco-BRL Life Technologies, New England Biolabs, Pharmacia, Stratagene.

Magnetic beads (streptavidin conjugate): Dynal Inc., Dynal (UK) Ltd.

Oligonucleotides: see 'Custom oligonucleotide synthesis' above.

PCR product purification: Hybaid, Perkin-Elmer Corp., Promega, Qiagen, Stratagene.

Phosphoramidites and associated reagents: Applied Biosystems, Cruachem.

Radioisotopes: Amersham, ICN.

Thermal cyclers: Appligene, Hybaid, Perkin-Elmer Corp., M J Research Inc., Techne.

Thermostable DNA polymerases: Amersham, Appligene, Boehringer Mannheim (Diagnostics and Biochemicals), New England Biolabs, Perkin-Elmer Corp., Promega.

Ultraviolet products (transilluminators, safety equipment): Amersham, Anachem, Appligene.

Addresses

Amersham Life Science, Inc., 2636 South Clearbrook Drive, Arlington Heights, IL 60005, USA. Tel: 708 593 6300; fax: 708 593 8010.

Amersham International plc, Amersham Place, Little Chalfont, Buckinghamshire HP7 9NA, UK. Tel: 0800 515313; fax: 0800 616927; http://www.amersham.co.uk/life

Anachem Ltd, Anachem House, 20 Charles Street, Luton, Bedfordshire LU2 0EB, UK. Tel: 01582 456666; fax: 01582 391768.

Applied Biosystems (a division of Perkin-Elmer), 850 Lincoln Center Drive, Foster City, CA 94404-1128, USA. Tel: 415 570 6667; fax: 415 572 2743.

Applied Biosystems Ltd (a division of Perkin-Elmer), Kelvin Close, Birchwood Science Park North, Warrington, Cheshire WA3 7PB, UK. Tel: 01925 825650; fax: 01925 282502.

Appligene Oncor, Pinetree Centre, Durham Road, Birtley, Chester-le-Street, Durham DH3 2TD, UK. Tel: 0191 492 0022; fax: 0191 492 0617.

Bio-Rad Life Science Group, 2000 Alfred Nobel Drive, Hercules, CA 94547, USA. Tel: 510 741 1000; fax: 510 741 1060.

Bio-Rad Laboratories Ltd, Bio-Rad House, Maylands Avenue, Hemel Hempstead, Hertfordshire HP2 7TD, UK. Tel: 0800 181134; fax: 01442 259118.

Boehringer Mannheim UK (Diagnostics and Biochemicals) Ltd, Bell Lane, Lewes, East Sussex BN7 1LG, UK. Tel: 0800 521578; fax: 0800 181087; http://biochem.boehringer-mannheim.com/

Cambridge BioScience, 24–25 Signet Court, Newmarket Road, Cambridge CB5 8LA, UK. Tel: 01223 316855; fax: 01223 360732. http://www.bioscience.co.uk

Clontech Laboratories Inc., 1020 East Meadow Circle, Palo Alto, CA 94303-4607, USA. Tel: 415 424 8222; fax: 415 424 1064; http://www.clonetech.com. UK Distributor: Cambridge BioScience.

Corning Costar Ltd, 1 The Valley Centre, Gordon Road, High Wycombe, Buckinghamshire HP13 6EQ, UK. Tel 01494 684700; fax: 01494 464891.

Cruachem Ltd, Todd Campus, West of Scotland Science Park, Acre Road, Glasgow G20 0UA, UK. Tel: 0141 945 0055; fax: 0141 946 0066.

Cruachem Inc., 45150 Business Court, Suite 550, Dulles, VA 20166, USA. Tel: 800 327 9362/703 689 3390; fax: 800 654 8646/703 689 3392.

Dynal Inc., 5 Delaware Drive, Lake Success, NY 11042, USA. Tel: 800 638 9416: fax: 516 326 3298.

Dynal (UK) Ltd, 10 Thursby Road, Croft Business Park, Bromborough, Wirral L62 3PW, UK. Tel: 0151 346 1234; fax: 0151 346 1223; e-mail: techserv@dynal.u-net.com

FMC BioProducts, 191 Thomaston Street, Rockland, ME 04841, USA. Tel: 207 594 3400; fax: 207 594 3426. UK distributor: Flowgen Instruments, Shenstone, Lichfield, Staffordshire WS14 0EE, UK. Tel: 01543 483054; fax: 01543 483055; e-mail: support@flowgen.philipharris.co.uk

GATC GmbH, Fritz Arnold Str. 23, D-78467 Konstanz, Germany, Tel: 75 31 81 60 0; fax 75 31 81 60 81; e-mail: service@gatc.de.

Gibco-BRL: *see* Life Technologies

Gilson Inc., Box 620027, 3000 West Beltline Highway, Middleton, WI 53562-0027, USA. Tel: 608 836 1551; fax: 608 831 4451; e-mail: sales@gilson.com. UK agents: Anachem Ltd, Anachem House, 20 Charles Street, Luton, Bedfordshire LU2 0EB, UK. Tel: 01582 456666; fax: 01582 391768.

Hybaid: *see* Life Sciences International

ICN Pharmaceuticals Inc., 3300 Hyland Avenue, Costa Mesa, CA 92626, USA. Tel: 800 854 0530; fax: 800 334 6999; http://www.icnpharm.com

ICN Biomedicals Ltd, Unit 18, Thame Park Business Centre, Wenman Road, Thame, Oxfordshire OX9 3XA, UK. Tel: 01844 213366; fax: 01844 213399.

Invitrogen Corporation, 1600 Faraday Avenue, Carlsbad, CA 92008, USA. Tel: 800 955 6288; fax: 760 603 7201; http://www.invitrogen.com

Invitrogen BV, De Schelp 12, 9351 NV Leek, The Netherlands. Tel: 594 515175; fax: 594 515312; http://www.invitrogen.com

Life Technologies Corp., Industrial Bioproducts, PO Box 6009, Gaithersburg, MD 20884-9980, USA. Tel: 301 840 8000; fax: 716 774 6639.

Life Technologies Ltd, 3 Fountain Drive, Inchinnan Business Park, Paisley PA4 9RF, UK. Tel: 0141 814 6100; fax: 0141 814 6287; http://www.lifetech.com

M J Research Inc., 149 Grove Street, Watertown, MA 02172, USA. Tel: 800 735 8437; fax: 617 923 8080.

MWG-Biotech, Anzunger Straße 7, D-85560 Ebersberg, Germany. Tel: 80 92 210 84; fax: 80 92 82 89 77; http://www.mwgdna.com/biotech

New England Biolabs, 32 Tozer Road, Beverley, MA 01915-5599, USA. Tel: 800 632 5227; fax: 508 921 1350; http://www.uk.neb.com

New England Biolabs (UK) Ltd, 67 Knowl Piece, Wilbury Way, Hitchin, Hertfordshire SG4 0TY, UK. Tel: 01462 420616; fax: 01462 421057; http://www.neb.com

Perkin-Elmer Corp., Applied Biosystems Division, 850 Lincoln Center Drive, Foster City, CA 94404-1128, USA. Tel: 415 570 6667; fax: 415 572 2743.

Perkin-Elmer Applied Biosystems, Kelvin Close, Birchwood Science Park North, Warrington, Cheshire WA3 7PB, UK. Tel: 01925 825650; fax: 01925 282502.

Pharmacia LKB Biotechnology Inc., 800 Centennial Avenue, PO Box 1327, Piscataway, NJ 08855-1327, USA. Tel: 800 526 3593; fax: 800 329 3593.

Pharmacia Biotech Ltd, 23 Grosvenor Road, St Albans, Hertfordshire AL1 3AW, UK. Tel: 01727 814000; fax: 01727 814001; http://www. biotech.pharmacia.se

Promega Corporation, 2800 Woods Hollow Road, Madison, WI 53711-5399, USA. Tel: 800 356 9526; fax: 608 277 2516.

Promega UK Ltd, Delta House, Enterprise Road, Chilworth Research Centre, Southampton SO1 7NS, UK. Tel: 0800 760225; fax: 01703 767014; http://www.euro.promega.com/uk.html

Qiagen GmbH, Max-Volmer-straße 4, 40724 Hilden, Germany. Tel: 2103 892230; fax; 2103 892222.

Qiagen Inc., 9600 De Soto Avenue, Chatsworth, CA 91311, USA. Tel: 800 425 8157; fax: 818 718 2056.

Qiagen Ltd, Boundary Court, Gatwick Road, Crawley, West Sussex RH10 2AX, UK. Tel: 01293 422900; fax: 01293 422922.

Stratagene, 11011 North Torrey Pines Road, La Jolla, CA 92037, USA. Tel: (800) 424 5444; fax: (619) 535 5400.

Stratagene Ltd, 140 Cambridge Science Park, Milton Road, Cambridge CB4 4GF, UK. Tel: 01223 420955; fax: 01223 420234.

Techne Inc, University Park Plaza, 743 Alexander Road, Princeton, NJ 08540-6328, USA. Tel: 609 452 9257; fax: 609 987 8177.

Techne (Cambridge) Ltd, Duxford, Cambridge CB2 4PZ, UK. Tel: 01223 832401; fax: 01223 836838; e-mail: sales@techneuk.attmail.com

Index